# Mathematical Modeling and Scale-up of Liquid Chromatography

# Springer

*Berlin*
*Heidelberg*
*New York*
*Barcelona*
*Budapest*
*Hong Kong*
*London*
*Milan*
*Paris*
*Tokyo*

Tingyue Gu

# Mathematical Modeling and Scale-up of Liquid Chromatography

With 75 Figures

 Springer

Prof. Dr. Tingyue Gu
Ohio University
Dept. of Chemical Engineering
Stocker Center
Athens, Ohio 45701-2979
USA

ISBN 3-540-58884-1 Springer-Verlag Berlin Heidelberg New York

CIP-data applied for

© Springer-Verlag Berlin Heidelberg 1995
Printed in Germany

The use of general descriptive names, registered names, trademarks, etc. in this publication does not imply, even in the absence of a specific statement, that such names are exempt from the relevant protective laws and regulations and therefore free for general use.

Production: PRODUserv Springer Produktions-Gesellschaft, Berlin
Cover-layout: Design & Production, Heidelberg
Typesetting: Dataconversion by Satztechnik Neuruppin GmbH, Neuruppin

SPIN 10100713    02/3020- 5 4 3 2 1 0    Printed on acid-free paper

# Preface

Liquid chromatography is no longer limited to chemical analysis. It has become an indispensable tool for the preparative- and large-scale purifications of proteins and other fine chemicals. So far, the scale-up of liquid chromatography relies mostly on trial and error and a few scaling rules that are more of a rule-of-thumb nature.

This book provides numerical solutions to a series of general multi-component rate models for liquid chromatography. The models consider dispersion, interfacial film mass transfer, intraparticle diffusion, and nonlinear multicomponent isotherm, or the second order kinetics. The models can be used to simulate various chromatographic operations. They provide more realistic descriptions of preparative- and large-scale liquid chromatography than the equilibrium theory and plate models because various mass transfer mechanisms are included.

The applications of the Fortran 77 codes for the models are explained. Parameter estimation for the models is discussed. The codes should be helpful in both the understanding of the dynamics of liquid chromatography and its scale-up. The codes are available to readers upon request by a letter, or preferably an electronic mail (to guting@ent.ohiou.edu).

Most of this book is based on the theoretical part of the author's Ph.D. thesis work at Purdue University, West Lafayette, Indiana, U.S.A. I am deeply indebted to my former advisor, Prof. George T. Tsao of the Laboratory of Renewable Resources Engineering at Purdue.

Spring 1995                                                            Tingyue Gu

# Table of Contents

# List of Symbols and Abbreviations

| | |
|---|---|
| $a_i$ | constant in Langmuir isotherm for component $i$, $b_i \, C_i^\infty$ |
| $b_i$ | adsorption equilibrium constant for component $i$, $k_{ai} / k_{di}$ |
| $Bi_i$ | Biot number of mass transfer for component $i$, $k_i R_p / (\varepsilon_p D_{pi})$ |
| $C_{0i}$ | concentration used for nondimensionalization, $\max\{C_{fi}(t)\}$ |
| $C_{bi}$ | bulk-fluid phase concentration of component $i$ |
| $C_{fi}$ | feed concentration profile of component $i$, a time dependent variable |
| $C_i$ | mobile phase concentration of component $i$ in ion-exchange |
| $C_{pi}$ | concentration of component $i$ in the stagnant fluid phase inside particle macropores |
| $C_p^c$ | critical concentration for concentration crossover in a binary isotherm |
| $C_{pi}^*$ | concentration of component $i$ in the solid phase of particle (based on unit volume of particle skeleton) |
| $C_i^\infty$ | adsorption saturation capacity for component $i$ (based on unit volume of particle skeleton) |
| $c_{bi}$ | $= C_{bi}/C_{0i}$ |
| $c_{pi}$ | $= C_{pi}/C_{0i}$ |
| $c_{pi}^*$ | $= C_{pi}^*/C_{0i}$ |
| $c_i^\infty$ | $= C_i^\infty/C_{0i}$ |
| $\underline{C}$ | saturation capacity in ion-exchange |
| $\overline{C}_i$ | stationary phase concentration of component $i$ in ion-exchange |
| $D_{bi}$ | axial or radial dispersion coefficient of component $i$ |
| $D_m$ | molecular diffusivity |
| $D_{pi}$ | effective diffusivity of component $i$, porosity not included |
| $Da_i^a$ | Damkölher number for adsorption, $L(k_{ai}C_{0i})/v$ |
| $Da_i^d$ | Damkölher number for desorption, $L(k_{di})/v$ |
| $d$ | inner diameter of a column |
| $F_i^{ex}$ | size exclusion factor for component $i$ ($F_i^{ex}=0$ means complete exclusion), $\varepsilon_{pi}^a / \varepsilon_p$ |
| $k_i$ | film mass transfer coefficient of component $i$ |
| $k_{ai}$ | adsorption rate constant for component $i$ |
| $k_{di}$ | desorption rate constant for component $i$ |
| $k_i'$ | retention factor (capacity factor) for component $i$ |
| $L$ | column length |

| $N$ | number of interior collocation points |
|---|---|
| Ne | number of quadratic elements |
| Ns | number of components |
| $Pe_{Li}$ | Peclet number of axial dispersion for component $i$, $vL/D_{bi}$ |
| $Q$ | mobile phase volumetric flow rate |
| $R$ | radial coordinate for particle |
| $R_p$ | particle radius |
| Re | Reynolds number, $v\rho_f(2R_p)/\mu$ |
| $r$ | $=R/R_p$ |
| Sc | Schmidt number, $\mu/(\rho_f D_m)$ |
| Sh | Sherwood number, $k(2R_p)/D_m$ |
| t | dimensional time ($t=0$ is the moment a sample enters a column) |
| $t_0$ | dead volume time of unretained small molecules, such as salts and solvents |
| $t_d$ | dead volume time of unretained large molecules, such as blue dextrin |
| $t_R$ | dimensional retention time |
| $v$ | interstitial velocity, $4Q/(\pi d^2 \varepsilon_b)$ |
| $W$ | weight of adsorbent |
| $Z$ | axial coordinate |
| $z$ | dimensionless axial coordinate, $Z/L$ |

*Greek Letters*

| $\alpha$ | $=2\sqrt{V+V_0}\left(\sqrt{1+V_0}-\sqrt{V_0}\right)$ for radial flow chromatography (RFC) |
|---|---|
| $\alpha_{ij}$ | separation factor of components $i$ and $j$ in ion-exchange |
| $\alpha_i, \beta_i, \gamma_i$ | parameters for the eluite-modulator correlation |
| $\varepsilon_b$ | bed void volume fraction |
| $\varepsilon_p$ | particle porosity |
| $\eta_i$ | dimensionless constant, $\varepsilon_p D_{pi} L/(R_p^2 v)$ |
| $\theta_{ij}$ | discount factors for extended multicomponent Langmuir isotherm |
| $\mu$ | mobile phase viscosity |
| $\xi_i$ | dimensionless constant for component $i$, $3Bi_i\eta_i(1-\varepsilon_b)/\varepsilon_b$ |
| $\rho_f$ | mobile phase density |
| $\rho_p$ | particle density, based on unit volume of particle skeleton |
| $\tau$ | dimensionless time, $vt/L$ |
| $\tau_R$ | dimensionless retention time |
| $\tau_{imp}$ | dimensionless time duration for a rectangular pulse of the sample |
| $\phi$ | phase ratio (stationary phase to mobile phase), $(1-\varepsilon_b)(1-\varepsilon_p)/[\varepsilon_b+(1-\varepsilon_b)\varepsilon_p]$ |

*Subscripts*

| a | adsorption reaction |
|---|---|
| b | bulk-fluid phase |
| d | desorption reaction |

| i | i-th component |
| imp | sample impulse |
| L | bulk-fluid phase |
| p | particle phase |

## Superscripts

| a | adsorption |
| c | isotherm crossover concentration point |
| d | desorption |
| * | particle phase concentration |
| $\infty$ | saturation capacity |

# 1 Introduction

High Performance Liquid Chromatography (HPLC) is undoubtedly one of the most important tools in chemical analysis. It has become increasingly popular at preparative- and large-scales, especially in purifying proteins. At such scales, larger particles are often used to pack the columns in order to reduce column pressure and facilitate column packing. Unlike small scale analytical HPLC columns that may give near plug flow performances, in large HPLC and lower pressure liquid chromatographic columns, dispersion and mass transfer effects are often important.

At smaller scales, the scale-up of liquid chromatography columns can usually be carried out by trial and error. Success depends to a large extent on the experience of the researcher. A failure is often not detrimental. The situation changes where large-scale columns are concerned. Such columns may easily cost thousands of dollars. There is much at stake in scale-up. A wrong estimation will render a purchased, or self-manufactured column, unsuitable for a particular application because of either insufficient resolution or low sample loading capacity. When an appropriate mathematical model is applied, it can be very helpful in supplementing the researcher's experience during scale-up.

There are several kinds of mathematical models for liquid chromatography with different complexities. A brief review of the models is given in Chap. 2. The equilibrium theory and plate models are generally not adequate for the realistic modeling and scale-up of multicomponent liquid chromatography, because of their inability to detail mass transfer mechanisms involved in preparative- and large-scale chromatographic separations.

The comprehensive mathematical models for liquid chromatography are often called the general multicomponent rate models, since they consider axial dispersion, interfacial mass transfer between the mobile and the stationary phases, intraparticle diffusion, and multicomponent isotherms. They provide an attractive alternative to the equilibrium theory and plate models for the modeling and scale-up of multicomponent chromatography. In the past, such a model was difficult to solve numerically on a computer. Due to tremendous advances in computer hardware, the model can now be solved using a common minicomputer. Because the general rate model considers different mass transfer mechanisms in a column, it is suitable for the realistic modeling of preparative- and large-scale chromatography. Computer simulation using the model also provides an excellent tool for studying many chromatographic phenomena without doing actual experiments.

In Chap. 3 of this book, a robust and efficient numerical procedure is presented to solve a general nonlinear multicomponent rate model that considers axial dispersion, external film mass transfer, intraparticle diffusion and complicated nonlinear isotherms. It uses quadratic finite elements for the discretization of the bulk-fluid phase partial differential equation (PDE) and the orthogonal collocation method for the particle phase PDE in the model, respectively. The resulting ordinary differential equation (ODE) system is solved by Gear's stiff method [1]. The model has been extended to include second order kinetics and the size exclusion effect. An alternative boundary condition at the column exit to the Danckwerts boundary condition is also analyzed.

The theoretical study of mass transfer in liquid chromatography in Chap. 4 points out the effects of mass transfer on multicomponent chromatographic separations. The influence of mass transfer related dimensionless parameters in the general multicomponent rate model is demonstrated by simulation. Also shown is an interesting case in which a component with an unfavorable isotherm does not show the expected anti-Langmuir peak shape with a diffused front and sharpened tail. Instead, it gives a peak with a tail more diffused than the peak front because of slow mass transfer rates.

In Chap. 5, a unified approach to a better understanding of multicomponent interference effects under mass transfer conditions is proposed. It has been shown that a displacement effect can be used to explain the dominating interference effects arising from the competition for binding sites among different components in multicomponent chromatography. It has been concluded that the concentration profile of a component usually becomes sharper due to the displacement effect from another component, while the concentration front of the displacer is often diffused as a consequence. Five factors stemming from equilibrium isotherms, which tend to escalate the displacement effect in multicomponent elutions, have been investigated. They have important implications for interference effects in multicomponent elutions under column-overload conditions.

In multicomponent elutions, competing modifiers are sometimes added to the mobile phase to compete with sample solutes for binding sites in order to reduce the retention times of strongly retained sample solutes [2]. Peaks in the chromatogram corresponding to a modifier are called system peaks [3]. Studies of system peaks may provide useful information on the effect of modifiers on the sample solutes and interpretation of some chromatograms. In Chap. 6, system peaks are studied systematically using the general multicomponent rate model. Systems peak patterns have been summarized for binary elutions with one competing modifier in the mobile phase involving samples that are either prepared in the mobile phase or in an inert solution. Binary elutions with two competing modifiers have also been investigated briefly.

A methodology is presented in Chap. 7 for the development of kinetic and isotherm models for multicomponent adsorption systems with uneven saturation capacities for different components, which are either physically induced or due to chiral discrimination of binding sites. The extended multicomponent Langmuir isotherm derived with this methodology, which is thermodynamically consistent, has been used successfully to explain isotherm crossovers and to demonstrate the peak reversal phenomenon under column-overload conditions.

In Chap. 8, the kinetic and mass transfer effects are discussed. The rate-limiting step in chromatography is investigated. The general multicomponent rate model has been modified to account for a reaction in the fluid phase between macromolecules and soluble ligands for the study of affinity chromatography. The adsorption, wash and elution stages in affinity chromatography have been simulated and analyzed.

Chap. 9 presents a general rate model for multicomponent gradient elution. The semi-empirical relationship between the modulator concentration and eluite affinity with the stationary phase developed by Melander et al. [4] is used in the model. Examples of multicomponent elution with linear and nonlinear gradients are demonstrated using computer simulation.

A general multicomponent rate model for radial flow chromatography (RFC) has been solved in Chap. 10 using the same numerical approach as that for conventional axial flow chromatography (AFC). The radial dispersion and external film mass transfer coefficients are treated as variables in the model for AFC. The extension of the general multicomponent rate model for RFC to include second order kinetics, the size exclusion effect and liquid phase reaction for the study of affinity RFC is also mentioned.

In Chap. 11, methods and correlations useful for the estimation of mass transfer and isotherm data are reviewed. The general approach and practical considerations for the scale-up of liquid chromatography using the general multicomponent rate models have been discussed.

Fortran 77 computer codes for the numerical solutions to all the rate models discussed in this book are available to the reader by sending a letter or preferably e-mail (guting@ent.ohiou.edu) to the author. They can be used with a Unix computer or a high-end IBM-compatible personal computer with sufficient RAM. The International Mathematical and Statistical Libraries (IMSL) from IMSL, Inc. (Houston, Texas) is required for the codes. The usage of the codes is demonstrated where they are first introduced in their respective chapters. If the IMSL is not available, the user may find a substitute from a public domain on the Internet for non-commercial applications.

# 2 Literature Review

## 2.1 Theories for Nonlinear Multicomponent Liquid Chromatography

Many researchers have contributed to the modeling of liquid chromatography. There exist a dozen or more theories of different complexities. A comprehensive review on the dynamics and mathematical modeling of isothermal adsorption and chromatography has been given by Ruthven [5] who classified models into three general categories: equilibrium theory, plate models, and rate models.

### 2.1.1 Equilibrium Theory

According to Ruthven, the equilibrium theory of multicomponent isothermal adsorption was first developed by Glueckauf [6]. The interference theory by Helfferich and Klein [7] that is mainly aimed at stoichiometric ion-exchange systems with constant separation factors, and the mathematically parallel treatise for systems with multicomponent Langmuir isotherms by Rhee and coworkers [8, 9] are both extensions of the equilibrium theory.

Equilibrium theory assumes a direct local equilibrium between the mobile phase and the stationary phase, neglecting axial dispersion and mass transfer resistance. It effectively predicts experimental retention times for chromatographic columns with fast mass transfer rates. It provides general locations, or retention times of elution peaks, but it fails to describe peak shapes accurately if mass transfer effects are significant. Equilibrium theory has been used for the study of multicomponent interference effects [7] and the ideal displacement development [9]. Many practical applications have been reported [7, 10-14].

### 2.1.2 Plate Models

Generally speaking, there are two kinds of plate models. One is directly analogous to the tanks in series model for nonideal flow systems [5]. In such a model, the column is divided into a series of small artificial cells, each with complete mixing. This gives a set of first order ODEs that describe the adsorption and interfacial mass transfer. Many researchers have contributed to this kind of plate model [5, 15-17]. However, plate models of this kind are generally not suitable for multicomponent chromatography since the equilibrium stages may not be assumed equal for different components.

The other kind of plate model is formulated based on the distribution factors that determine the equilibrium of each component in each artificial stage. The model solution involves recursive iterations rather than solving ODE systems. The most popular are the Craig distribution models. By considering the so-called blockage effect, the Craig models are applicable to multicomponent systems. Descriptions of Craig models were given by Eble et al. [18], Seshadri and Deming [19], and Solms et al. [20]. The Craig models have been used for the study of column-overload problems [18, 21]. Recently, Velayudhan and Ladisch [22] used a Craig model with a corrected plate count to simulate elutions and frontal adsorptions.

### 2.1.3 Rate Models

Rate models refer to models containing a rate expression, or rate equation, which describes the interfacial mass transfer between the mobile phase and the stationary phase. A rate model usually consists of two sets of differential mass balance equations, one for the bulk-fluid phase, the other for the particle phase. Different rate models have varying complexities [5].

#### 2.1.3.1 Rate Expressions

The solid film resistance hypothesis was first proposed by Glueckauf and Coates [23]. It assumes a linear driving force between the equilibrium concentrations in the stationary phase (determined from the isotherm) and the average fictitious concentrations in the stationary phase. This simple rate expression has been used by many researchers [5, 24-26] because of its simplicity, but this model cannot provide details of the mass transfer processes.

The fluid film mass transfer mechanism with a linear driving force is also widely used [5]. The driving force is the concentration difference of a component between that on the surface of a particle and that in the surrounding bulk-fluid. It is assumed that there is a stagnant fluid film between the particle surface and the bulk-fluid. The fluid film exerts a mass transfer resistance between the bulk-fluid phase and the particle phase, often called the external mass transfer resistance. If the concentration gradient inside the particle phase is ignored, the chromatography model then becomes a lumped particle model, which has been used by some researchers [27-29]. If the mass transfer Biot number, which reflects the ratio of the characteristic rate of film mass transfer to that of intraparticle diffusion, is much larger than one, the external film mass transfer resistance can be neglected with respect to intraparticle diffusion.

In many cases, both the external mass transfer and the intraparticle diffusion must be considered. A local equilibrium is often assumed between the concentration in the stagnant fluid phase inside macropores and the solid phase of the particle. Such a rate mechanism is adequate to describe the adsorption and mass transfer between the bulk-fluid and particle phases, and inside the particle phase in most chromatographic processes. The local equilibrium assumption here is different from that made for the equilibrium theory. The equilibrium theory assumes a direct equilibrium of concentrations in the solid and the liquid phases without any mass transfer resistance.

If the adsorption and desorption rates are not sufficiently high, the local equilibrium assumption is no longer valid. A kinetic model must be used. Some kinetic models were reviewed by Ruthven [5] and Lee et al. [30, 31]. Second order kinetics has been widely used in kinetic models for affinity chromatography [32-39].

### 2.1.3.2  Governing Equation for the Bulk-Fluid Phase

The governing partial differential equation for the bulk-fluid phase can be easily obtained from a differential mass balance of the bulk-fluid phase for each component. Axial dispersion, convection, transient, and the interfacial flux terms are usually included. Such equations themselves are generally linear if physical parameters are not concentration dependent. They become nonlinear when coupled with a rate expression involving nonlinear isotherms or second order kinetics.

For some rate models, such as models for isothermal, single component systems with linear isotherms, analytical solutions may be obtained using the Laplace transform [5, 31]. For more complex systems, especially those involving nonlinear isotherms, analytical solutions generally cannot be derived [5]. Numerical methods must be used to obtain solutions to complex rate models that consider various forms of mass transfer mechanisms [40]. Detailed rate models are becoming increasingly popular, especially in the study of preparative- and large-scale chromatography.

### 2.1.3.3  General Multicomponent Rate Models

A rate model that considers axial dispersion, external mass transfer, intraparticle diffusion and nonlinear isotherms, is called a general multicomponent rate model. Such a model is adequate in most cases to describe the adsorption and mass transfer processes in multicomponent chromatography.

In some cases, surface adsorption, size exclusion and adsorption kinetics may have to be included to give an adequate description of a particular system. Several groups of researchers have used different numerical procedures to solve various general multicomponent rate models [40-43].

#### 2.1.3.4 Numerical Solutions

A general multicomponent rate model consists of a coupled PDE system with two sets of mass balance equations, one for the bulk-fluid and one for the particle phases for each component, respectively. The finite difference method is a simple numerical procedure that can be directly applied to the solution of the entire model [42, 44]. This often requires a huge amount of computer memory during computation, and its efficiency and accuracy are not competitive compared with other more advanced numerical methods, such as the orthogonal collocation (OC), finite element, or the orthogonal collocation on finite element (OCFE) methods.

For the particle phase governing equation, the OC method is the obvious choice. It is a very accurate, efficient and simple method for discretization. It has been widely used with success for many particle problems [45, 46]. The formulation of the OC method for particles is readily available from Finlayson's book [46].

Unfortunately, concentration gradients in the bulk-fluid phase can be very steep, thus the OC method is no longer a desirable choice, since global splines using high order polynomials are very expensive [46] and sometimes unstable. The method of OCFE uses linear finite elements for global splines and collocation points inside each element. No numerical integration for element matrices is needed because of the use of linear elements. This discretization method can be used for systems with stiff gradients [46].

The finite element method with higher order of interpolation functions (typically quadratic, or occasionally cubic) is a very powerful method for stiff systems. Its highly streamlined structure provides unsurpassed convenience and versatility. This method is especially useful for systems with variable physical parameters, as in radial flow chromatography and nonisothermal adsorption with or without chemical reactions. Chromatography of some biopolymers also involves a variable axial dispersion coefficient [47].

#### 2.1.3.5 Solution to the ODE System

If the finite element method is used for the discretization of the bulk-fluid phase PDE and the OC method for the particle phase equations, an ODE system is produced. The ODE system with initial values can be readily solved using an ODE solver such as subroutine "IVPAG" of the IMSL [48] software package, which uses the powerful Gear's stiff method [40].

## 2.2 Scale-Up of Liquid Chromatography

Currently, scale-up of liquid chromatography is carried out largely based on trial-and-error and experience, with the help of some general scale-up rules that are not necessarily accurate. Some of these rules were discussed by Snyder et al. [2], and Pieri et al. [49]. They are mostly empirical or semi-empirical relationships about particle size, flow rate, column length, and resolution. The correlations are more of a "rule of thumb" nature when they are used for scale-up. Knox and Pyper [50] did an extensive study on column-overload. Some of their results on concentration and volume overload are also helpful in the scale-up of liquid chromatography. There are many papers in this area. They are not the focus of this book.

Instead of following these scale-up rules, a rate model can be used to simulate chromatograms of a larger column before it is built or purchased. The model uses only a few experimental data from a small column with the same packing as a large column. Although rate models have great potential in more accurate scale-up of liquid chromatography, most applications have been on the simulation of smaller columns to match experimental chromatograms. Practical examples involving larger columns have not been reported in the literature.

# 3 A General Multicomponent Rate Model for Column Liquid Chromatography

## 3.1 Model Assumptions

For the modeling of multicomponent liquid chromatography, the column is divided into the bulk-fluid phase and the particle phase. The anatomy of a fixed-bed axial flow chromatography column is given in Fig. 3.1. To formulate a general rate model, the following basic assumptions are required.

(1) The chromatographic process is isothermal. There is no temperature change during a run.

**Bulk-Fluid Phase**

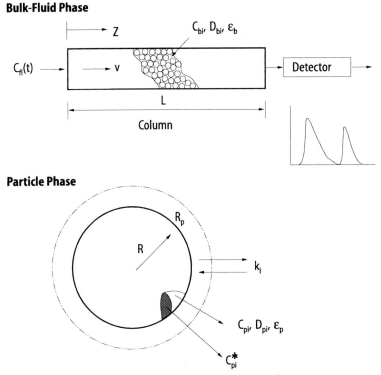

**Particle Phase**

**Fig. 3.1.** Modeling of fixed-bed axial flow chromatography

(2) The porous particles in the column are spherical and uniform in diameter.
(3) The concentration gradients in the radial direction are negligible.
(4) The fluid inside particle macropores is stagnant, i.e., there is no convective flow inside macropores.
(5) An instantaneous local equilibrium exists between the macropore surfaces and the stagnant fluid inside macropores of the particles.
(6) The film mass transfer mechanism can be used to describe the interfacial mass transfer between the bulk-fluid and particle phases.
(7) The diffusional and mass transfer parameters are constant and independent of the mixing effects of the components involved.

## 3.2 Model Formulation

Based on the assumptions above, the governing equations can be obtained from differential mass balances of the bulk-fluid phase and the particle phase, respectively, for component $i$. The following equations can also be derived from equations of continuity provided by Bird et al. [51]:

$$-D_{bi}\frac{\partial^2 C_{bi}}{\partial Z^2}+v\frac{\partial C_{bi}}{\partial Z}+\frac{\partial C_{bi}}{\partial t}+\frac{3k_i(1-\varepsilon_b)}{\varepsilon_b R_p}\left(C_{bi}-C_{pi,R=R_p}\right)=0 \qquad (3\text{-}1)$$

$$\left(1-\varepsilon_p\right)\frac{\partial C_{pi}^*}{\partial t}+\varepsilon_p\frac{\partial C_{pi}}{\partial t}-\varepsilon_p D_{pi}\left[\frac{1}{R^2}\frac{\partial}{\partial R}\left(R^2\frac{\partial C_{pi}}{\partial R}\right)\right]=0 \qquad (3\text{-}2)$$

with the initial and boundary conditions

$$t=0, \quad C_{bi}=C_{bi}(0,Z); \quad C_{pi}=C_{pi}(0,R,Z) \qquad (3\text{-}3,4)$$

$$Z=0, \quad \frac{\partial C_{bi}}{\partial Z}=\frac{v}{D_{bi}}\left(C_{bi}-C_{fi}(t)\right); \quad Z=L, \quad \frac{\partial C_{bi}}{\partial Z}=0 \qquad (3\text{-}5,6)$$

$$R=0, \quad \frac{\partial C_{pi}}{\partial R}=0; \quad R=R_p, \quad \frac{\partial C_{pi}}{\partial R}=\frac{k_i}{\varepsilon_p D_{pi}}\left(C_{bi}-C_{pi,R=R_p}\right) \quad . \qquad (3\text{-}7,8)$$

Defining the following dimensionless constants,

$$c_{bi}=C_{bi}/C_{0i}, \quad c_{pi}=C_{pi}/C_{0i}, \quad c_{pi}^*=C_{pi}^*/C_{0i}, \quad \tau=vt/L, \quad r=R/R_p$$

$$z=Z/L, \quad \text{Pe}_{Li}=vL/D_{bi}, \quad \text{Bi}_i=k_i R_p/(\varepsilon_p D_{pi}), \quad \eta_i=\varepsilon_p D_{pi}L/(R_p^2 v)$$

$$\xi_i=3\text{Bi}_i\eta_i(1-\varepsilon_b)/\varepsilon_b$$

the model equations can be transformed into the following dimensionless equations:

$$-\frac{1}{\text{Pe}_{Li}}\frac{\partial^2 c_{bi}}{\partial z^2}+\frac{\partial c_{bi}}{\partial z}+\frac{\partial c_{bi}}{\partial \tau}+\xi_i\left(c_{bi}-c_{pi,r=1}\right)=0 \qquad (3\text{-}9)$$

$$\frac{\partial}{\partial \tau}\left[\left(1-\varepsilon_p\right)c_{pi}^* + \varepsilon_p c_{pi}\right] - \eta_i\left[\frac{1}{r^2}\frac{\partial}{\partial r}\left(r^2\frac{\partial c_{pi}}{\partial r}\right)\right] = 0 \quad . \tag{3-10}$$

Initial conditions:

$$\tau = 0, \quad c_{bi} = c_{bi}(0,z); \quad c_{pi} = c_{pi}(0,r,z) \tag{3-11,12}$$

Boundary conditions:

$$z = 0, \quad \frac{\partial c_{bi}}{\partial z} = \text{Pe}_{Li}\left[c_{bi} - \frac{C_{fi}(\tau)}{C_{0i}}\right] \tag{3-13}$$

where $C_{0i}$ is the concentration used to nondimensionalize other concentrations for component $i$. It should be the highest concentration of component $i$ ever fed to the column, i.e., $C_{0i} = \max\{C_{fi}(t)\}$ with $-\infty < t < +\infty$ .

For frontal adsorption, $\quad\quad\quad\quad\quad C_{fi}(\tau)/C_{0i} = 1$

For elution, $\quad\quad\quad\quad\quad\quad\quad\quad C_{fi}(\tau)/C_{0i} = \begin{cases} 1 & 0 \le \tau \le \tau_{imp} \\ 0 & \text{else} \end{cases}$ .

After the sample introduction (in the form of frontal adsorption):

If component $i$ is displaced, $\quad\quad C_{fi}(\tau)/C_{0i} = 0$

If component $i$ is a displacer, $\quad\quad C_{fi}(\tau)/C_{0i} = 1$ .

At $z=1$, $\partial c_{bi}/\partial z = 0$ $\hfill$ (3-14)

At $r=0$, $\partial c_{pi}/\partial r = 0$; at $r=1$, $\partial c_{pi}/\partial r = \text{Bi}_i(c_{bi} - c_{pi,r=1})$ . $\hfill$ (3-15,16)

The definitions of all the dimensionless variables and parameters are listed in the list of Symbols and Abbreviations. In Eq. (3-10), $c_{pi}^*$ is the dimensionless concentration of component $i$ in the solid phase of the particles. It is directly linked to a multicomponent isotherm such as the following commonly used multicomponent Langmuir isotherm:

$$C_{pi}^* = \frac{a_i C_{pi}}{1 + \sum_{j=1}^{Ns} b_j C_{pj}}, \quad \text{i.e.,} \quad c_{pi}^* = \frac{a_i c_{pi}}{1 + \sum_{j=1}^{Ns}\left(b_j C_{0j}\right)c_{pj}} \quad \text{(dimensionless)} \quad . \tag{3-17}$$

The stoichiometric isotherm with constant separation factors used by Helfferich and Klein [7] for ion-exchange can be written as follows,

$$\overline{C}_i = \frac{\overline{C}C_i}{\sum_{j=1}^{Ns}\alpha_{ji}C_j} = \frac{\alpha_{1,Ns}\overline{C}C_i}{\sum_{j=1}^{Ns}\alpha_{j,Ns}C_j} \tag{3-18}$$

in which $\alpha_{ij} = 1/\alpha_{ji} = \alpha_{ik}\alpha_{kj}$, and $\alpha_{ii} = 1$. $C_i$ is the concentration of ion component $i$ in the mobile phase. $\overline{C}$ is the saturation capacity and is considered equal for all components. $\overline{C}_i$ is the concentration of ion component $i$ in the stationary phase. All the concentrations in Eq. (3-18) are based on the unit volume of the column rather than of the respective phases as in the case of Langmuir isotherms.

The stoichiometric isotherm for ion-exchange can be converted into an isotherm shown in Eq. (3-19), which is the same in algebraic expression as the Langmuir isotherm, except that "1+" in the denominator in the Langmuir isotherm expression is dropped.

$$C_{pi}^* = \frac{a_i C_{pi}}{\sum_{j=1}^{Ns} b_j C_{pj}}, \quad \text{i.e.,} \quad c_{pi}^* = \frac{(a_i / C_{0i}) c_{pi}}{\sum_{j=1}^{Ns} b_j c_{pj} C_{0j} / C_{0j}} . \tag{3-19}$$

The following relationships are needed for the conversion:

$$b_i = \alpha_{i,Ns} \quad \text{and} \quad a_i = b_i C^\infty = \frac{\alpha_{i,Ns} \overline{C}}{(1 - \varepsilon_b)(1 - \varepsilon_p)} \quad (i = 1,2,\ldots,Ns) \tag{3-20a,b}$$

where ion component Ns is assigned as the basis of the separation factors. Note that the units of $a_i$ and $b_i$ in the Langmuir isotherm and the converted stoichiometric isotherm are different. In the stoichiometric isotherm, the concentrations of components cannot all be zero at the same time, which means that the column is never "empty." This is not so for the Langmuir isotherm.

If the general rate model is solved, it provides the effluent history (chromatogram), $c_{bi}|_{z=1}$ vs $\tau$. In fact the model even provides transient concentration profiles anywhere inside the column, either in the bulk-fluid $[c_{bi}(r,z)]$, or the stagnant fluid inside particle macropores, or in the solid skeleton of the particles $[c_{bi}^*(\tau,r,z)]$. Usually only the effluent history is used to study chromatographic phenomena. The transient concentration profiles in the bulk-fluid phase inside the column provides information on peak migration. It may be useful in some theoretical studies.

The model itself is not necessarily nonlinear, but it becomes nonlinear whenever a nonlinear isotherm, such as the Langmuir isotherm, is used. A true multicomponent case is almost certainly nonlinear, since no linear isotherm can be used to describe true multicomponent adsorptions. For such a nonlinear multicomponent model, there is no analytical solution. The model equations must be solved numerically. Figure 3.2 shows the strategy of the numerical method used to solve the PDE system in the model. The bulk-fluid phase and the particle phase equations are first discretized using the FE and the OC methods, respectively. The resulting ODE system is solved using an existing ODE solver that is based on the Gear's stiff method.

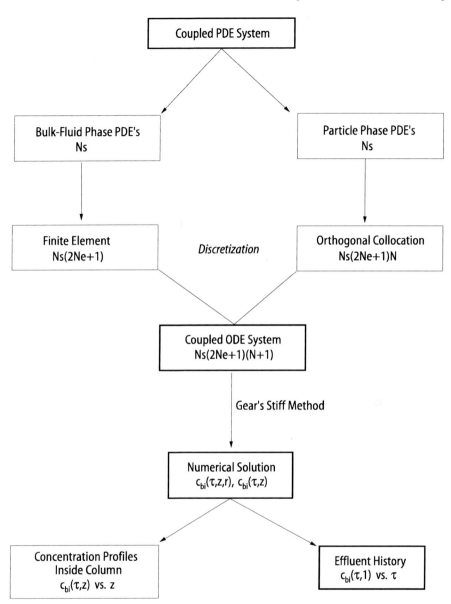

**Fig. 3.2.** Solution strategy

## 3.3 Finite Element Formulation for the Bulk-Fluid Phase Governing Equation

Applying the Galerkin weighted residual method [52] to Eq. (3.9) one obtains

$$\int_{z_A}^{z_B} \phi_m \left[ -\frac{1}{Pe_{Li}} \frac{\partial^2 c_{bi}}{\partial z^2} + \frac{\partial c_{bi}}{\partial z} + \frac{\partial c_{bi}}{\partial \tau} + \xi_i \left( c_{bi} - c_{pi,r=1} \right) \right] dz = 0 \tag{3-21}$$

where the integration limit $\{z_A, z_B\}$ contains the two boundary points of a typical finite element. Rearrangement using integration by parts on the second order partial differential term in Eq. (3-21) gives,

$$\int_{z_A}^{z_B} \frac{1}{Pe_{Li}} \frac{\partial c_{bi}}{\partial z} \frac{\partial \phi_m}{\partial z} dz + \left( \frac{-1}{Pe_{Li}} \right) \phi_m \frac{\partial c_{bi}}{\partial z} \bigg|_{z_A}^{z_B} + \int_{z_A}^{z_B} \phi_m \frac{\partial c_{bi}}{\partial \tau} dz +$$

$$+ \int_{z_A}^{z_B} \left( \phi_m \frac{\partial c_{bi}}{\partial z} + \xi_i \phi_m c_{bi} \right) dz - \int_{z_A}^{z_B} \xi_i \phi_m c_{pi,r=1} \, dz = 0 \quad . \tag{3-22}$$

Inserting the interpolation form for quadratic elements $c_{bi} = \sum_{n=1}^{3} \phi_n c_{bi,n}$ into Eq. (3-22) yields

$$\sum_{n=1}^{3} c_{bi,n} \int_{z_A}^{z_B} \frac{1}{Pe_{Li}} \frac{\partial \phi_m}{\partial z} \frac{\partial \phi_n}{\partial z} dz + \sum_{n=1}^{3} c_{bi,n} \int_{z_A}^{z_B} \left( \phi_m \frac{\partial \phi_n}{\partial z} + \xi_i \phi_m \phi_n \right) dz +$$

$$+ \sum_{n=1}^{3} c'_{bi,n} \int_{z_A}^{z_B} \phi_m \phi_n dz = \left( PB_i \right)_m + \int_{z_A}^{z_B} \xi_i \phi_m c_{pi,r=1} dz \tag{3-23}$$

where $\left( PB_i \right)_m = \left( -\frac{1}{Pe_{Li}} \right) \phi_m \frac{\partial c_{bi}}{\partial z} \bigg|_{z_A}^{z_B}$ .

Equation (3-23) can be expressed in the matrix form as follows:

$$[DB_i][c'_{bi}] + [AKB_i] = [PB_i] + [AFB_i] \tag{3-24}$$

where the bold face indicates a matrix or a vector for each quadratic element, and

$$\left( DB_i \right)_{m,n} = \int_{z_A}^{z_B} \phi_m \phi_n \, dz \tag{3-25}$$

$$\left( AKB_i \right)_{m,n} = \int_{z_A}^{z_B} \left( \frac{1}{Pe_{Li}} \frac{\partial \phi_m}{\partial z} \frac{\partial \phi_n}{\partial z} + \phi_m \frac{\partial \phi_n}{\partial z} + \xi_i \phi_m \phi_n \right) dz \tag{3-27}$$

$$(\mathbf{AFB}_i)_{m,n} = \int\limits_{z_A}^{z_B} \xi_i \phi_m c_{pi,r=1}\, dz \tag{3-26}$$

in which $m,n \in \{1,2,3\}$. The finite element matrices and vectors are evaluated over each individual element before a global assembly. After the global assembly, the natural boundary condition $(\mathbf{PB}_i)|_{z=0} = -c_{bi} + C_{fi}(\tau)/C_{0i}$ will be applied to $[\mathbf{AKB}_i]$ and $[\mathbf{AFB}_i]$ at $z = 0$. $(\mathbf{PB}_i) = 0$ anywhere else.

## 3.4  Orthogonal Collocation Formulation of the Particle Phase Governing Equation

Using the same symmetric polynomials as defined by Finlayson [46], Eq. (3-10) is transformed into the following equation by the OC method:

$$\left(\sum_{j=1}^{Ns} \frac{\partial g_i}{\partial c_{pj}} \frac{dc_{pj}}{d\tau}\right)_l = \eta_i \sum_{k=1}^{N+1} B_{l,k}(c_{pi})_k\,, \quad l = 1,2,\ldots,N \tag{3-28}$$

in which $g_i = (1-\varepsilon_p)c_{pi}^* + \varepsilon_p c_{pi}$. Note that for component $i$, $c_{pi}^*$ is related to $c_{pj}$ values for all the components involved via multicomponent isotherms such as Eq. (3-17). The value of $(c_{pi})_{N+1}$, i.e., $c_{pi,r=1}$, can be obtained from the boundary condition at $r=1$, which gives

$$\sum_{j=1}^{N+1} A_{N+1,j}(c_{pi})_j = Bi_i(c_{bi} - c_{pi,r=1}) \tag{3-29}$$

or

$$c_{pi,r=1} = \frac{Bi_i c_{bi} - \sum\limits_{j=1}^{N} A_{N+1,j}(c_{pi})_j}{A_{N+1,N+1} + Bi_i} \,. \tag{3-30}$$

In Eqs. (3-29) and (3-30) the matrices $\mathbf{A}$ and $\mathbf{B}$ are the same as defined by Finlayson [46], who also provided their values in his book.

## 3.5  Solution to the ODE System

If Ne quadratic elements [i.e., (2Ne+1) nodes] are used for the $z$-axis in the bulk-fluid phase equation and N interior OC points are used for the $r$-axis in the particle phase equation, the above discretization procedure gives a total of

Ns(2Ne+1)(N+1) ODEs that are then solved simultaneously by Gear's stiff method using the IMSL subroutine "IVPAG" [48]. A function subroutine must be supplied to the ODE solver IVPAG to evaluate concentration derivatives at each element node and interior OC point with given trial concentration values. The concentration derivatives at each element node $[c'_{bi}]$ are determined from Eq. (3-24). The concentration derivatives at each OC point $[c'_{bi}]$ are coupled because of the complexity of the isotherms that are related to $g_i$ via $c^*_{pi}$ in multicomponent cases. At each interior OC point, Eq. (3-28) can be rewritten in the matrix form shown in Eq. (3-31):

$$[GP][c'_p] = [RH] \qquad (3\text{-}31)$$

where $GP_{ij} = \partial g_i / \partial c_{pj}$, $c'_{pj} = dc_{pj}/d\tau$, and $[RH_i]$ = right hand side of Eq. (3-28). Since the matrix $[GP]$ and the vector $[RH]$ are known with given trial concentration values at each interior OC point in the function subroutine, the vector $[c'_p]$ can be easily calculated from Eq. (3-31).

## 3.6  Fortran 77 Code for the General Multicomponent Rate Model

A Fortran 77 code has been written for the numerical solution to the general multicomponent rate model. The code is named "RATE.F." The Unix computer or IBM-compatible Personal Computer (PC) used to run the Fortran 77 code must have the IMSL library already installed. To run this code on a PC, it is necessary to use a high-end PC with a math coprocessor and sufficient RAM. In this book, all simulations were obtained on Unix minicomputers, such as the SUN 4/280 minicomputer, or SUN 4/390 Sparc workstation.

Double precision is used for the Fortran 77 code. It was found that using double precision instead of single precision does not increase CPU (Central Processing Unit) time significantly. Concentration values of $c_{bi}$ and $c_{pi}$ are stored in the vector named "**u**". The vector "**u**" is divided first according to the sequence of components. It is then subdivided into $c_{bi}$ values and $c_{pi}$ values for all element nodes on z-axis. The $c_{pi}$ values are subdivided according to the sequence of interior collocation points for particles. The exterior collocation point (N+1) is not stored in "**u**". Normally, only $c_{bi}$ values at the column exit at different dimensionless time $\tau$ are printed out. These values can be used directly to plot a simulated effluent history. If a user wants the concentration profiles inside the bulk-fluid phase in the column, or the particle phase concentrations, he/she may rearrange the printout statements in the code. Such data are three-dimensional for a single component, and they take a lot of space to print out.

The core of the Fortran 77 code is subroutine "fcn." It supplies concentration derivatives for the ODE solver IVPAG. Readers are urged to consult the IMSL Manual (Version 1.0) [48] in order to understand how this IVPAG is used. The error tolerance for IVPAG is set to tol=$10^{-5}$ throughout this book. Although IVPAG allows a choice between the Gear's stiff method and the Adam's method, the former is chosen because of its ability to deal with stiff ODE systems. The MAIN program of the code does the preparation of calling IVPAG, and input and output operations. The finite element notations used in the code generally follow those used in the textbook by Reddy [52].

After compilation using the command "f77 RATE.F -limslib" on a Unix computer, an executable file named "a.out" will be generated. The "-limslib" sign indicates that the code is linked to the IMSL library during the compilation. A separate data file named "data" must be supplied when running the code. When "a.out" is executed, it reads the file "data" automatically for input data.

Each chromatographic operation (or mode) is assigned an "index" value that is included in the file "data." The following is a list of operations allowed in the code.

index = 1,  breakthrough curve

index = 2,  isocratic elution with an inert mobile phase (containing no modifier)

index = 3,  step-change displacement. The last component is displacer

index = 4,  breakthrough switched to displacement at t = tshift

index = 5,  same as index = 4, but the flow direction is reversed for displacement

index = 6,  isocratic elution with a modifier in the mobile phase (component Ns). The sample contains the same modifier concentration as in the mobile phase

index = 7,  same as index = 6, but sample is in an inert solution.

If the user desires other forms of operation, he may change the initial conditions in the MAIN program, and the Natural Boundary Condition (NBC) in subroutine "fcn" in the code accordingly. The NBC is linked to the column feed profiles of all components. Such changes are quite easy to make.

If the stoichiometric isotherm with separation factors for ion-exchange are used instead of the built-in Langmuir isotherm, the index value is increased by 10. For example, index = 13 indicates a step-change displacement operation involving components with stoichiometric isotherms. Also, in the input file, $a_i$ and $b_i$ values should be those obtained according to their relationships with separation factors shown in Eqs. (3-20a,b).

The input data for the Fortran 77 code contains the number of components, elements and interior collocation points, respectively, and process index, time control data, dimensionless parameters, isotherm type and parameters. Note that the code is based on the dimensionless PDE systems and $C_{0i}$ can be combined with $b_i$ to form a dimensionless group, $b_i C_{0i}$. This does not mean that increasing the $C_{0i}$ value has the same effect as increasing the $b_i$ value, since the latter also increases the $a_i$ value proportionally.

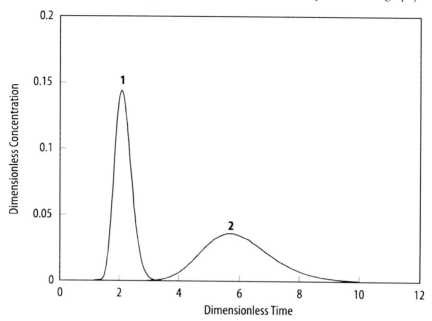

**Fig. 3.3.** Simulation of a binary elution using the rate model

Below is the input file "data" used to obtain Fig. 3.3 that shows a binary elution with an inert mobile phase. The numbers in file "data" can be separated by a space (or spaces), or a comma, or a new line:

```
2  7  2  2  0.1 0.03 10 0.4 0.5
  300  4  20  0.1  1.2  1.5
  300  4  20  0.1  8.0  10.0    .
```

There is no designated output file for the code. The user may use the command "a.out > results" to save the output into a file named "results." The file "results" for Fig. 3.3 has the following heading:

```
Multicomponent Chromatography Simulator by Tingyue Gu (Ohio U.)
================================================================
nsp nelemb nc index  timp   tint  tmax  epsip  epsib
 2    7    2   2    0.100  0.030  10.0  0.400  0.500
_____

   PeL      eta      Bi    CO       consta   constb
  300.00   4.000  20.000  0.10000   1.200    1.500
  300.00   4.000  20.000  0.10000   8.000    10.000
================================================= End of data file
Total ODE = 90   data pts = 333
_____

index =1 Breakthru; =2, Elution with inert MP
index =3 Step-change disp. Last comp. is displacer
index =4 BT, switch to displacement at t = tshift
```

```
index =5 Same as index=4, but reverse flow
index =6 Elution, the last component is modifier
index =7 Same as =6, but sample is in inert...
index =10+ use separation factors
```

```
Results (t, c1, c2, ...) follow.  Please wait...
```

```
    0.0300  0.00000  0.00000
    (... more data points)     .
```

The numerical numbers in the heading before the words "End of data file" are the data read from the input file "data" by the code upon its execution. They are the number of components (nsp), the number of elements (nelemb), the number of interior collocation points (nc), the operation index, $\tau_{imp}$ (timp), the dimensionless time interval in the output data (tint), the maximum $\tau$ for calculation (tmax), $\varepsilon_p$ (epsip) , $\varepsilon_b$ (epsib) , $Pe_{Li}$ (PeL) , $\eta_i$ (eta), $B_i$ (Bi), the maximum concentration $C_{0i}$ (C0) for each component, $a_i$ (consta), and $b_i$ (constb). Note that $a_i / b_i$ (moles per unit volume of particle skeleton) value for all the components must be the same in order to have the same saturation capacity $(C^\infty)$ for all the components.

If a $\tau_{imp}$ value is not needed for a chromatographic operation, such as breakthrough operation, an arbitrary value, say 1.0, is still assigned to it in the file "data" to maintain the data structure. The code will read this value, but it is not used in calculation. If index = 4 or 5, an additional dimensionless shift time must be provided at the end of file "data." This time value indicates when the column's feed is switched to the displacer after the rest of the components have been going through a frontal adsorption stage. The displacer is listed as the last component in file "data." A "tmax" value is required to tell the code to stop calculation after $\tau$ reaches this value. A "tint" value is needed to control the data points in the output file by specifying a time interval. This and "tmax" together determine how many data points are there in the output file. In this book, all the simulated chromatograms are plotted on a PC using Lotus Freelance Software (Lotus Corporation, Cambridge, MA) by linearly linking data points from the output files, which usually have 200 to 500 points. The three dimensionless parameters used in the data input are calculated from the following relationships based on their definition: $Bi_i = k_i R_p / (\varepsilon_p D_{pi})$, $Pe_{Li} = vL / D_{bi}$, and $\eta_i = \varepsilon_p D_{pi} L / (R_p^2 v)$.

If the user wants to use an isotherm other than the multicomponent Langmuir isotherm or the stoichiometric ion-exchange isotherm, the subroutine named "getdgdc" in RATE.F has to be rewritten to provide $(\partial g_i / \partial c_{pi})$ values. $g_i = (1 - \varepsilon_p) c_{pi}^* + \varepsilon_p c_{pi}$, in which $c_{pi}^*$ is related to the isotherm. Of course, the input of the Fortran 77 code may also have to be modified to receive the new isotherm parameters. All the modifications should be straightforward.

The so-called "equilibrium-dispersive model" used by Guiochon et al. [27] is a subcase of the general rate model. The former is actually a lumped particle model with the Langmuir isotherm that is discussed in the next section. The "equilibrium-dispersive model" is accommodated by the Fortran 77 code

RATE.F. The user only has to set the particle porosity to zero in the input data file. Note that the "equilibrium-dispersive model" lumps the intraparticle diffusion into the axial dispersion and film mass transfer resistance, thus the pseudo-axial dispersion and pseudo-film mass transfer coefficients should not be evaluated using the existing correlations in the literature (see Chap. 11) during parameter estimation. Otherwise, the "lumping" effect is not there.

## 3.7   CPU Time for the Simulation

Effluent concentration profiles can be obtained from the numerical solution to the model. The model also provides the effluent history and the moving concentration profiles inside the column for each component. The concentration profile of each component inside the stagnant fluid phase and the solid phase of the particle can also be obtained, but they are rarely used for discussions.

The CPU (Central Processing Unit) time required for the simulation of a chromatogram depends largely on how many ODEs need to be solved. The total number of ODEs is equal to $Ns(2Ne+1)(N+1)$. Systems with more components and stiff concentration profiles require more ODEs to be solved.

Usually, two interior collocation points ($N=2$) are needed, especially when $D_{pi}$ values are small, which in turn give large $Bi_i$ and small $\eta_i$ values. Sometimes one interior collocation point ($N=1$) is sufficient. The value of $N$ does not affect the stability of the numerical solution. For theoretical studies, using $N=1$ saves a lot of time while achieving the same qualitative results. If the model is used for data fitting, the error generated by using $N=1$ instead of $N=2$ will probably be less than the experimental error. Insufficient $N$ tends to give diffused concentration profiles as shown in Figs. 3.4-3.6. Using $N=1$ instead of $N=2$ in Figs. 3.4-3.6 (dashed lines) saves about 60% CPU time on a SUN 4/280 computer.

The selection of number of elements, $Ne$, is quite important. There is no exact a priori criterion to dictate an $Ne$ value, although the general rule is that the stiffer the concentration profiles the higher the $Ne$ value. An insufficient $Ne$ value will result in a numerical solution that has oscillation. $Ne=5-10$ is usually sufficient for systems with nonstiff or slightly stiff concentration profiles. For stiff cases, $Ne=20-30$ is often enough. A small $Ne$ value can be tried out first. If the solution shows oscillation, a rerun can be carried out with an increased $Ne$ value. In Fig. 3.6, the dotted lines are obtained by using three quadratic finite elements ($Ne=3$) and one interior collocation point ($N=1$) with a CPU time of only 13.2 s. Though the dotted lines show a certain degree of oscillation, they still provide the general shape of the converged concentration profiles, which take 6.9 min of CPU time to calculate. This means that one may use small $Ne$ and $N$ values to get the rough concentration profiles very quickly and then decide what to do next.

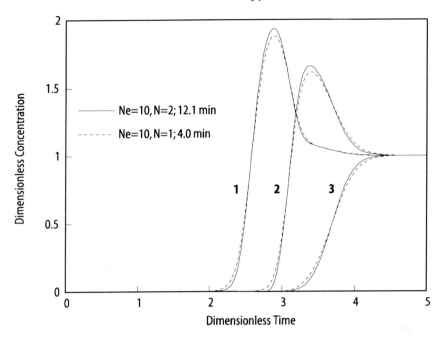

**Fig. 3.4.** Effect of the number of interior OC points in the simulation of frontal adsorption

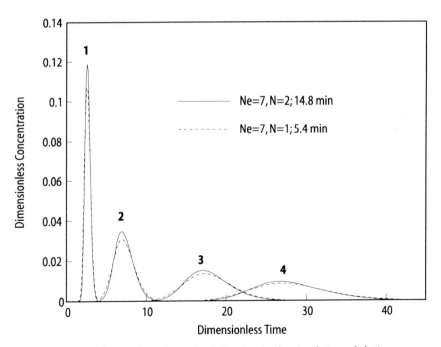

**Fig. 3.5.** Effect of the number of interior OC points in the simulation of elution

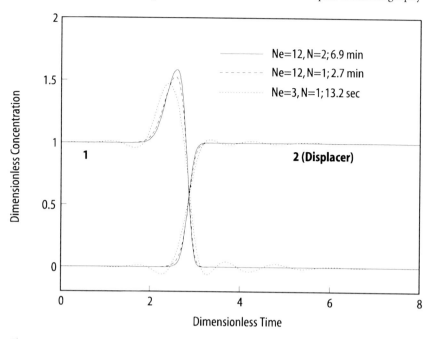

**Fig. 3.6.** Convergence of the concentration profiles of a stepwise displacement system

In Chap. 4, it will be shown that large $Pe_{Li}$, $Bi_i$ and $\eta_i$ values contributes to stiff concentration profiles. Isotherm parameters may also have some effects. For example, if the sample concentration falls into the nonlinear range of the Langmuir isotherm during elution, the self-sharpening effect causes a very stiff front-flank of the peak. In theoretical studies it is unnecessary to run very stiff cases unless required. Most arguments can be made based on simulations of less stiff concentration profiles without demanding excessive computing resources. The improvement in computer hardware will allow the user to run practically any reasonable simulation in the near future.

## 3.8  Extension of the General Multicomponent Rate Model

The general multicomponent rate model can be extended to include second order kinetics in order to describe a system with slow binding/dissociation. It can also be extended to include the size exclusion effect.

### 3.8.1 Second Order Kinetics

In the basic multicomponent rate model, it is assumed that there exists a local equilibrium for each component between the stagnant fluid phase inside macropores and the solid phase of the particles. This assumption may not be satisfied if the adsorption and desorption rates are not high, or the mass transfer rates are relatively much faster. In such cases, isotherm expressions cannot be directly inserted into Eq. (3-10) to replace $c_{pi}^*$. Instead, a second order kinetic expression can be used. It has been widely adopted to account for reaction kinetics in the study of affinity chromatography [32-38]. A rate model with second order kinetics was applied to affinity chromatography by Arve and Liapis [38].

Second order kinetics assumes the following common reversible binding and dissociation reaction:

$$P_i + L \underset{k_{di}}{\overset{k_{ai}}{\rightleftharpoons}} P_i L \tag{3-32}$$

where $P_i$ is component $i$ in the fluid and $L$ represents immobilized ligands. $k_{ai}$ and $k_{di}$ are the adsorption and desorption rate constants for component $i$, respectively. The binding kinetics is of second order and the dissociation first order. The rate equation for Eq. (4-32) is as follows,

$$\frac{\partial C_{pi}^*}{\partial t} = k_{ai} C_{pi} \left( C^\infty - \sum_{j=1}^{Ns} C_{pi}^* \right) - k_{di} C_{pi}^* \ . \tag{3-33}$$

The rate constant $k_{ai}$ has a unit of concentration over time while the rate constant $k_{di}$ has a unit of inverse time. If the reaction rates are relatively large compared to mass transfer rates, then instant adsorption/desorption equilibrium can be assumed such that both sides of Eq. (3-33) can be set to zero, which subsequently gives the Langmuir isotherm with the equilibrium constant $b_i = k_{ai} / k_{di}$ for each component.

Introducing dimensionless groups $Da_i^a = L(k_{ai} C_{0i})/v$ and $Da_i^a = L k_{di}/v$ that are defined as the Damkölher numbers [53] for adsorption and desorption, respectively, Eq. (3-33) can be nondimensionalized as follows:

$$\frac{\partial c_{pi}^*}{\partial \tau} = Da_i^a c_{pi} \left( c_i^\infty - \sum_{j=1}^{Ns} \frac{C_{0j}}{C_{0i}} c_{pj}^* \right) - Da_i^d c_{pi}^* \ . \tag{3-34}$$

If the saturation capacities are the same for all the components, at equilibrium, Eq. (3-34) gives $b_i C_{0i} = Da_i^a / Da_i^a$ and $a_i = C^\infty b_i = c_i^\infty Da_i^a / Da_i^d$ for the resultant multicomponent Langmuir isotherm.

Equation (3-34), which is an ODE, replaces the Langmuir isotherm and it does not complicate the numerical procedure for the solution of the model since the discretization process is untouched. One only has to add Eq. (3-34) in the final ODE system. The final ODE system consists of Eqs. (3-9), (3-10) and (3-34). With the trial values of $c_{bi}$, $c_{pi}$ and $c_{pi}^*$ in the function subroutine in the Fortran 77 code, their derivatives can easily be evaluated from the three ODE expressions. If Ne elements and N interior collocation points are used for the discretization of Eqs. (3-9) and (3-10), there will be Ns(2Ne+1)(2N+1) ODEs in the final ODE system, which is Ns(2Ne+1)N more than in the equilibrium case [54]. These extra ODEs come from Eq. (3-34) at each element node and each interior collocation point for each component.

### 3.8.2  Addition of Size Exclusion Effect to the Rate Model

Several mathematical models have been proposed for size exclusion chromatography [55-57]. The model proposed by Kim and Johnson [56] is similar to the general rate model described in this work, except that their model considers size exclusion in single component systems without any adsorption. They introduced an "accessible pore volume fraction" to account for the size exclusion effect.

In this book, a symbol $\varepsilon_{pi}^a$ is used to denote the accessible porosity (i.e., accessible macropore volume fraction) for component i. For small molecules with no size exclusion effect, $\varepsilon_{pi}^a = \varepsilon_p$, and for large molecules that are completely excluded from the particle macropores $\varepsilon_{pi}^a = 0$. For medium-sized molecules, $0 < \varepsilon_{pi}^a < \varepsilon_p$. It is convenient to define a size exclusion factor $0 \leq F_i^{ex} \leq 1$ such that $\varepsilon_{pi}^a = F_i^{ex} \varepsilon_p$. To include the size exclusion effect, Eq. (3-2) should be modified as follows,

$$\left(1 - \varepsilon_p\right)\frac{\partial C_{pi}^*}{\partial t} + \varepsilon_{pi}^a \frac{\partial C_{pi}}{\partial t} - \varepsilon_{pi}^a D_{pi}\left[\frac{1}{R^2}\frac{\partial}{\partial R}\left(R^2 \frac{\partial C_{pi}}{\partial R}\right)\right] = 0 \qquad (3\text{-}35)$$

where the first term, $(1-\varepsilon_p)\partial C_{pi}^* / \partial t$, should be dropped or set to zero if the component does not bind with the stationary phase. It should be pointed out again that $C_{pi}^*$ in Eq. (3-35) is based on the unit volume of particle skeleton (solid portion of the particles), excluding the volume of pores measured by the particle porosity $\varepsilon_p$. For a component that is completely excluded from the particles (i.e. $\varepsilon_{pi}^a = 0$), adsorbing only on the outer surface of the particles, Eq. (3-35) degenerates into the following interfacial mass balance relationship:

$$\frac{\partial C_{pi}^*}{\partial t} = \frac{3k_i}{(1-\varepsilon_p)R_p}\left(C_{bi} - C_{pi, R=R_p}\right) = 0 \qquad (3\text{-}36)$$

Equation (3-36) can be combined with the bulk-fluid phase governing equation, Eq. (3-1), to give the following equation that is similar to a lumped particle model:

$$-D_{bi}\frac{\partial^2 C_{bi}}{\partial Z^2}+v\frac{\partial C_{bi}}{\partial Z}+\frac{\partial C_{bi}}{\partial t}+\frac{(1-\varepsilon_b)(1-\varepsilon_p)}{\varepsilon_b}\frac{\partial C_{pi}^*}{\partial t}=0 \qquad (3\text{-}37)$$

where $C_{pi}^*$ either follows the multicomponent isotherm or the expression for second order kinetics. Eq. (3-38) is the dimensionless form of Eq. (3-37),

$$-\frac{1}{Pe_{Li}}\frac{\partial^2 c_{bi}}{\partial z^2}+\frac{\partial c_{bi}}{\partial z}+\frac{\partial c_{bi}}{\partial \tau}+\frac{(1-\varepsilon_b)(1-\varepsilon_p)}{\varepsilon_b}\frac{\partial c_{pi}^*}{\partial \tau}=0 \quad . \qquad (3\text{-}38)$$

If component $i$ does not bind with the stationary phase, $C_{pi}^*\equiv 0$ and the fourth term in the left-hand side of Eq. (3-37) is dropped for that component. If no component is totally excluded, the addition of the size exclusion effect in the rate models is very simple. One only has to use $\varepsilon_{pi}^a D_{pi}$ to replace $\varepsilon_p D_{pi}$ in the expression of $Bi_i$ and $\eta_i$, and $\varepsilon_{pi}^a$ in $\varepsilon_p c_{pi}$ of Eq. (3-10).

Mathematically, a singularity occurs in the model system when a component (say, component $i$) is totally excluded from the particles (i.e., $\varepsilon_{pi}^a=0$) if one does not use Eq. (3-38) to replace Eqs. (3-9) and (3-10). It turns out that, for numerical calculation, there is no need to worry about this singularity if $\varepsilon_{pi}^a$ is given a very small value below that of the tolerance of the ODE solver, which is set to $10^{-5}$ throughout this book. It is found through simulation that this treatment gives practically the same results as those obtained by using Eq. (3-38). Setting $Bi_i=\eta_i=(k_iL/R_pv)^{1/2}$ can help avoid oscillations.

One should be aware that the size exclusion effect of a component affects its saturation capacity in the isotherm. It also affects the effective diffusivity of the component since the tortuosity is related to the accessible porosity. Clearly, using size exclusion in a multicomponent model may lead to the use of uneven saturation capacities for a component with significant size exclusion effect and a component with no size exclusion effect. This may cause problems when the multicomponent Langmuir isotherm is used because the isotherm becomes thermodynamically inconsistent [5].

The Fortran 77 code for the rate model with second order kinetics and the size exclusion effect is named KINETIC.F. It works in a similar way to RATE.F. To explain the input file "data" for the code, the heading of the result file for Fig. 3.7 is shown below.

```
Chromatography Simulator with 2nd Order Kinetics
and Size Exclusion Effect by Tingyue Gu (Ohio U.)
================================================================
nsp nelemb nc index  timp   tint  tmax  epsip epsib
 2   16    2   2     0.400  0.040 16.0  0.400 0.400

   PeL      eta      Bi    Cinf     CO     Daa    Dad    exf
 300.00   4.000  10.000 0.50E+00 0.10E+01   4.000 40.000 0.900E+00
 300.00   4.000  10.000 0.50E+00 0.10E+011000.000 40.000 0.900E+00
===================================================== End of data file.
Total ODE = 330    Data pts = 400
```

```
index =1 Breakthru; =2, Elution with inert MP
index =3 Step-change disp. Last comp. is displacer
index =4 BT, switch to displacement at t = tshift
index =5 Same as index=4, but reverse flow
index =6 Elution, the last component is modifier
index =7 Same as =6, but sample is in inert...
```

```
Results (t, c1, c2, ... ) follow.    Please wait...
```

```
    0.0400   0.00000   0.00000
    (...more data points)    .
```

Stripping the text until reaching "End of data file" in the heading above, the remaining numerical figures are what the file "data" contains. In the printout above, "Cinf " is the saturation capacity $C_i^\infty$. "C0" is $C_{0i}$. Both dimensional concentrations should be based on moles, not the weight. "Daa" and "Dad" are $Da_i^a$ and $Da_i^d$, respectively. If both $Da_i^a$ and $Da_i^d$ values are large (say $10^3$-$10^5$ or larger), KINETIC.F will yield practically the same numerical results as RATE.F since second order kinetics degenerates into the local equilibrium assumption with the Langmuir isotherm. "exf" is the size exclusion factor $F_i^{ex}$. If there is no size exclusion factor, it should be set to unity. If there is total exclusion, its values should be set to a very small value (say $10^{-6}$) instead of zero as discussed previously. The computer program will automatically set its value to $10^{-5}$ if its input value is zero. When the Fortran 77

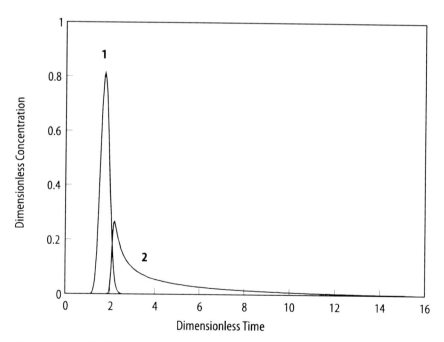

**Fig. 3.7.** Example of binary elution simulated using the kinetic model

code is used to show the effect of feed concentration, $C_{0i}$ (which is equal to $\max\{C_{fi}(t)\}$, i.e., the maximum concentration of component $i$ ever fed to the column) value changes. This also changes the value of $Da_i^a$ since it is proportional to $C_{0i}$ according to its definition.

## 3.9 The Question of Choosing Column Boundary Conditions

In this book, the Danckwerts boundary conditions [58] shown as Eqs. (3-5) and (3-6) are used for the two column ends. The validity of the Danckwerts boundary conditions in transient axial dispersion models has been discussed for many years by some researchers. Reviews were given by Parulekar [59], and Parulekar and Ramkrishna [60]. For axial dispersions in some linear systems, Parulekar and Ramkrishna [60] provided some physically more reasonable alternatives to the Danckwerts boundary conditions for transient systems based on analytical analyses. Unfortunately, for nonlinear systems, such an approach is generally not possible. Recently, Lee and coworkers [30, 31] discussed the use of alternative boundary conditions for both column inlet and exit in some rate models.

In nonlinear chromatography, the Danckwerts boundary conditions are generally accepted. However, for the column inlet some researchers [61, 62] implied that it is better to use finite values for the concentration flux instead of zero as in the Danckwerts boundary conditions. This is equivalent to assuming that the column is semi-infinitely long, and the effluent history is detected at $z=1$. This alternative boundary condition is hardly appropriate, since it tends to destroy the mass balance of the model system.

In the effluent history of a frontal analysis, each breakthrough curve can be integrated to see whether it matches the dimensionless column holdup capacity for the corresponding component, which is expressed by the following expression assuming that there is no size exclusion effect,

In Eq. (3-39), $CA_i$ consists of three parts, the amount of component $i$ adsorbed onto the solid part of the particles, that in the stagnant fluid inside particles, and that in the bulk-fluid:

$$CA_i = \frac{(1-\varepsilon_b)(1-\varepsilon_p)\dfrac{b_i C^\infty}{1+\displaystyle\sum_{j=1}^{Ns} b_j C_{0j}} + (1-\varepsilon_b)\varepsilon_p + \varepsilon_b}{\varepsilon_b}. \qquad (3\text{-}39)$$

Equation (3-39) is actually equal to the first moment of a breakthrough curve. It is equivalent to the expression of the first moment for a single component system with the Danckwerts boundary conditions and Langmuir isotherm derived by Lee et al. [63] from differential mass balance equations. The holdup capacity should also be equal to the area integrated from Eq. (3-40),

$$CA_i = \tau_e - \int_0^{\tau_e} c_{bi}\big|_{z=1}\, d\tau \tag{3-40}$$

where $\tau_e$ is a time value at which the breakthrough curve has already leveled off. Since the holdup capacity reflects the steady state of the column, mass transfer and dispersion effects should not affect its value. The above two equations are very helpful in checking the mass balance of an effluent history in frontal analysis and stepwise displacement. A Fortran 77 code named AREA.F is used for the calculation of the area above a concentration curve and below the dimensionless concentration = unity line, and the area under a concentration curve. It requires IMSL for compilation. Upon execution of the code, it asks for the name of your concentration profile data file that contains time and concentration data points in pairs for a single component. This code can similarly be used to calculate peak areas in elution.

The use of the Danckwerts boundary condition at the column exit needs no effort in the finite element formulation since, in the finite element method, the zero flux as a natural boundary condition is a default. The implementation of the alternative boundary condition at the column exit can also be easily accommodated in the existing code that uses the Danckwerts boundary conditions. In the function subroutine of the code that evaluates concentration derivatives, the trial concentration values are given; thus $(\partial c_{bi}/\partial z)\big|_{z=1}$ can be obtained by using the concentration values at the three element nodes for derivative calculations. In the actual code, adding a few lines and a simple extra subroutine for the derivative calculations will suffice for the modification needed. Note that the natural boundary condition at $z=1$ is

$$(\mathbf{PB}_i)\big|_{z=1} = \frac{1}{Pe_{Li}} \frac{\partial c_{bi}}{\partial z}\bigg|_{z=1} \tag{3-41}$$

and it should be added to $[\mathbf{AFB}_i]$ at $z=1$. Figure 3.8 shows single component breakthrough curves with the Danckwerts boundary condition and the alternative boundary condition at the column exit, respectively. Parameter values used for simulation are listed in Table 3.1. It is obvious that the use of the alternative boundary condition results in a later breakthrough, and thus a larger capacity area that is not equal to the correct theoretical value, unlike in the case with the Danckwerts boundary conditions. This violation of a basic mass balance is clearly undesirable. In fact, any attempt to change the Danckwerts boundary condition at one column end while leaving the other end intact may lead to such a violation since the complete Danckwerts boundary conditions are mass balanced.

Figure 3.9 shows the concentration profiles inside the column at different times. It shows that when the Danckwerts boundary condition is used at $z=1$, the concentration curves bend upward trying to approximate the zero flux requirement. Note that this boundary condition may not be completely satisfied with limited element nodes, but the $c_{bi}$ values at $z=1$ are easily converged. Figure 3.9 also shows that the concentration profiles are quite

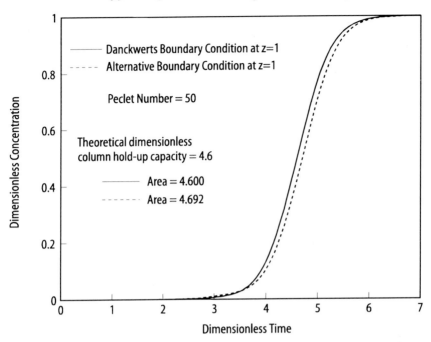

**Fig. 3.8.** Single-component breakthrough curves (Peclet number = 50)

**Table 3.1.** Parameter values used for simulation in Chap. 3[a]

| Figure(s) | Species | Physical Parameters | | | | | Numerical Parameters | |
|---|---|---|---|---|---|---|---|---|
| | | $Pe_{Li}$ | $\eta_i$ | $Bi_i$ | $a_i$ | $b_i \times C_{0i}$ | Ne | N |
| 3.3 | 1 | 300 | 4 | 20 | 1.2 | 1.5 ×0.1 | 7 | 2 |
| | 2 | 300 | 4 | 20 | 8 | 10 ×0.1 | | |
| 3.4 | 1 | 400 | 6 | 10 | 2 | 4 ×0.1 | | |
| | 2 | 400 | 6 | 10 | 7 | 12 ×0.1 | 10 | 2 |
| | 3 | 400 | 6 | 10 | 15 | 30 ×0.1 | | |
| 3.5 | 1 | 300 | 4 | 20 | 1.2 | 1.5 ×0.1 | 7 | 2 |
| | 2 | 320 | 4.2 | 17 | 8 | 10 ×0.1 | | |
| | 3 | 400 | 5.5 | 16 | 24 | 30 ×0.1 | | |
| | 4 | 500 | 7 | 15 | 38.4 | 48 ×0.1 | | |
| 3.6 | 1 | 600 | 6 | 5 | 3 | 6 ×0.1 | 12 | 2 |
| | 2 | 600 | 3 | 6 | 12 | 24 ×0.3 | | |
| 3.8 | 1 | 50 | 2 | 10 | 8 | 7 ×0.2 | 4 | 2 |
| 3.9 | 1 | 50 | 2 | 10 | 8 | 7 ×0.2 | 20 | 2 |
| 3.10 | 1 | 50 | 10 | 4 | 4 | 3.5 ×0.2 | 8 | 2 |
| | 2 | 50 | 10 | 4 | 8 | 7 ×0.2 | | |
| 3.11 | 1 | 200 | 2 | 10 | 8 | 7 ×0.2 | 5 | 2 |

[a] For Figs. 3.4 to 3.11, $\varepsilon_b = \varepsilon_p = 0.4$. Sample size for Fig. 3.5 is $\tau_{imp} = 0.1$; for Fig. 3.10, $\tau_{imp} = 0.2$.

smooth at the column exit when the alternative boundary condition is used. This is because the alternative boundary condition assumes that the column has no discontinuity at the column exit. In Fig. 3.9, when $\tau$ is not very small, the concentration shapes are very similar for different $\tau$. This is the so-called "constant pattern" phenomenon [5].

The mass balance violation may not occur or be noticeable in elution, as is shown by Fig. 3.10, in which the areas for each component for the Danckwerts boundary condition and the alternative boundary conditions at $z=1$ are both 0.2000, i.e., the value of sample size $\tau_{imp}$.

It is shown in Eq. (3-41) that the $(PB_i)|_{z=1}$ value can be set to zero if the $Pe_{Li}$ values are large. Fig. 3.11 has the same conditions as Fig. 3.8, except that the Peclet number in Fig. 3.11 is 200 that is much larger than that in Fig. 3.8. Fig. 3.11 shows that the differences are quite small from using the Danckwerts boundary condition or the alternative boundary condition at the column exit when the Peclet number is not small. This is in agreement with the results obtained by Brian et al. [61]. In fact, in common axial flow chromatography, the Peclet numbers for axial dispersion often run into the thousands or higher; the difference resulting from using different boundary conditions at the column exit is negligible. Thus, seeking an alternative boundary condition to replace the Danckwerts boundary condition for the column exit seems meaningless.

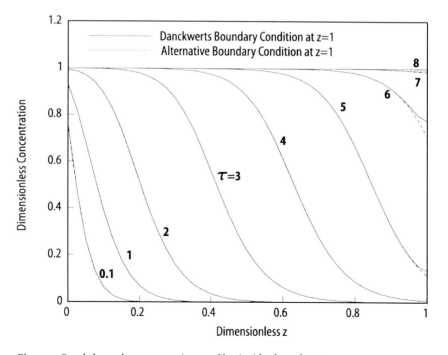

**Fig. 3.9.** Breakthrough concentration profiles inside the column

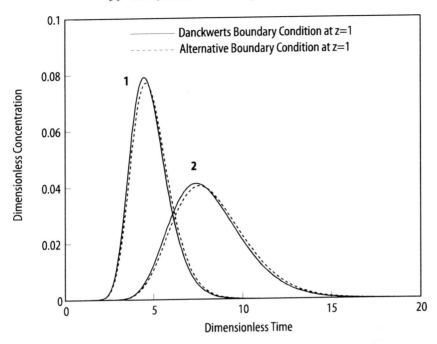

**Fig. 3.10.** Binary elution with different boundary conditions at the column exit

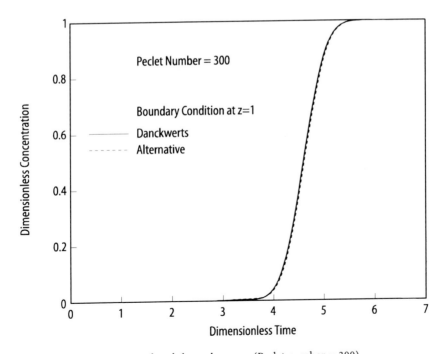

**Fig. 3.11.** Single-component breakthrough curves (Peclet number = 300)

# 4 Mass Transfer Effects

For analytical and some preparative columns in liquid chromatography, mass transfer resistance is usually negligible and the equilibrium theory suffices [7]. But for preparative columns with smaller plate numbers and large-scale columns, mass transfer effects are often significant and cannot usually be neglected.

The study of mass transfer effects for single component systems has been carried out by many researchers [5]. One group of researchers, Lee et al. [64] studied the mass transfer effects in nonlinear multicomponent elution ion-exchange chromatography. They compared the difference between a general rate model and the equilibrium theory under various mass transfer conditions.

## 4.1 Effects of Parameters $Pe_{Li}$, $Bi_i$ and $\eta_i$

The Peclet number ($Pe_{Li}$) reflects the ratio of the convection rate to the dispersion rate while the Biot number ($Bi_i$) reflects the ratio of the external film mass transfer rate to the intraparticle diffusion rate. Figures. 4.1 to 4.3 show that the increase of $Pe_{Li}$ values (while fixing other parameters) sharpens the concentration profiles in the effluent history in cases of frontal adsorption, elution and stepwise displacement. This well-known result has been reported by numerous researchers and summarized by Ruthven [5]. Parameter values used for simulation in this chapter are listed in Table 4.1. Figures. 4.4 to 4.6 show that increasing $\eta_i$ or $Bi_i$ also gives sharper concentration profiles. The figures show that increasing the dimensional parameter $k_i$ or $D_{pi}$ has the same effect. Clearly, mass transfer effects tend to diffuse concentration profiles.

## 4.2 Effect of Flow Rate

The volumetric mobile phase flow rate, $Q$, in an axial flow chromatography column is directly proportional to the interstitial velocity $v$. This velocity affects $D_{bi}$ and $k_i$ values. Meanwhile the intraparticle diffusivities, $D_{pi}$ can be regarded as independent of $v$ [65].

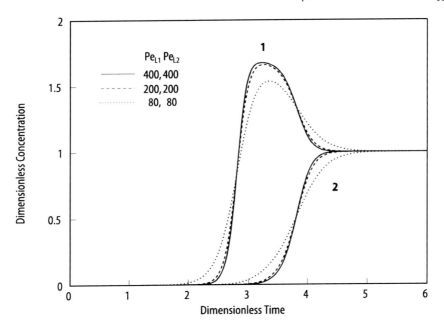

**Fig. 4.1.** Effect of Peclet numbers on two-component frontal adsorption

Dispersion is caused by molecular diffusion and turbulent mixing or eddy diffusion [5]. A simple linear approximation for a single component system may be represented by Ruthven [5] and Jonsson [66]:

$$D_b = \gamma_1 D_m + \gamma_2 (2R_p) v \tag{4-1}$$

where $\gamma_1$ and $\gamma_2$ are constants that normally have values of about 0.7 and 0.5, respectively. The molecular diffusion ($\gamma_1 D_m$) of a liquid is negligible com-

**Table 4.1.** Parameter values used for simulation in Chap. 4[a]

| Figure(s) | Species | Physical Parameters | | | | | Numerical Parameters | |
|---|---|---|---|---|---|---|---|---|
| | | $Pe_{Li}$ | $\eta_i$ | $Bi_i$ | $a_i$ | $b_i \times C_{0i}$ | Ne | N |
| 4.1–4.3 | 1 | 400 | 10 | 4 | 5 | $5 \times 0.2$ | 1 | 2 |
| | 2 | 400 | 10 | 4 | 20 | $20 \times 0.2$ | | |
| 4.4, 4.5 | 1 | 400 | | | 5 | $5 \times 0.2$ | 15 | 2 |
| | 2 | 400 | | | 20 | $20 \times 0.2$ | | |
| 4.6 | 1 | 400 | | | 5 | $5 \times 0.2$ | 15 | 2 |
| | 2 | 400 | | | 5 | $5 \times 0.4$ | | |
| 4.7 | 1 | 400 | | | 5 | $5 \times 0.2$ | 15 | 2 |
| | 2 | 400 | | | 20 | $20 \times 0.2$ | | |
| 4.8 | 1 | 300 | | | 4 | $10 \times 0.8$ | 5 | 2 |

[a] In all cases, $\varepsilon_b = \varepsilon_p = 0.4$. For elution cases the sample sizes are: $\tau_{imp} = 0.6$

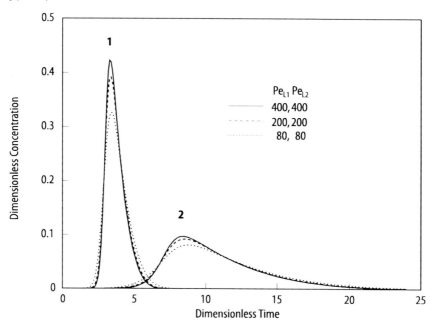

**Fig. 4.2.** Effect of Peclet numbers on two-component elution

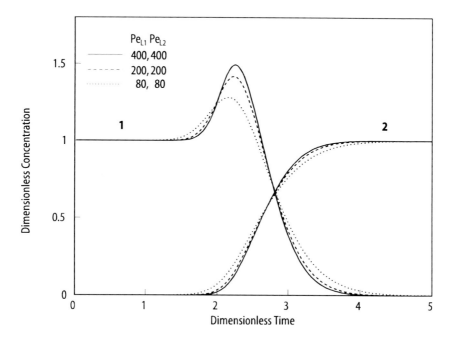

**Fig. 4.3.** Effect of Peclet numbers on two-component stepwise displacement

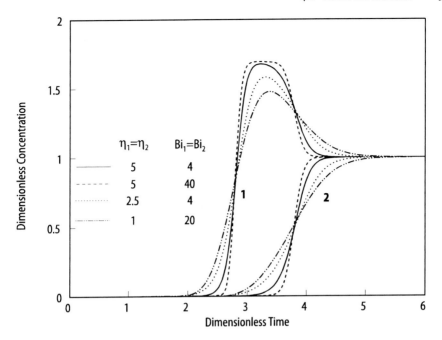

**Fig. 4.4.** Effect of $\eta_i$ and $Bi_i$ on binary frontal adsorption

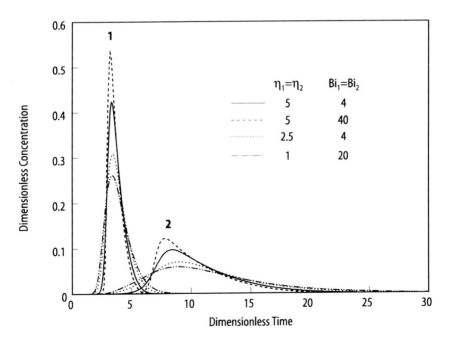

**Fig. 4.5.** Effect of $\eta_i$ and $Bi_i$ on binary elution

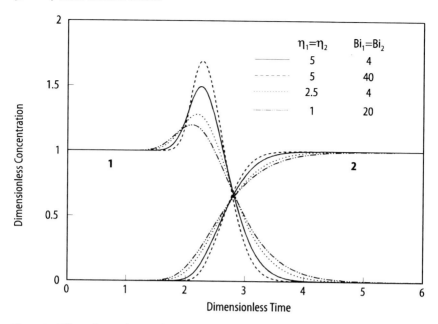

**Fig. 4.6.** Effect of $\eta_i$ and $Bi_i$ on binary stepwise displacement

pared to eddy diffusion, even at low Reynolds numbers [5]. In Eq. (4-1), eddy diffusivity is the dominant term in liquid chromatography, especially when the flow velocity is not low, thus $D_b \propto v$. This relationship has been acknowledged by some researchers [31, 67]. For simplicity in discussions, the multicomponent mixing effects on $D_{bi}$, $D_{pi}$ and $k_i$ for multicomponent systems are ignored in this book. Thus, $D_{bi} \propto v$ and $Pe_{Li}$ is independent of $v$ for each component.

The relationship between $k_i$ and $v$ can simply be expressed as $k_i \propto v^{1/3}$ [63]. It is in agreement with two different experimental correlations reported by Pfeffer and Happel [68], Wilson and Geankoplis [69] and Ruthven [5] for liquid systems at low Reynolds numbers ($Re = 2R_p v \rho / \mu$) that cover the range for liquid chromatography [70, 71]. From the relationship $k_i \propto v^{1/3}$, one obtains $Bi_i \propto k_i \propto v^{1/3}$.

Figure 4.7 clearly shows a case in which the sharpness and resolution of the elution peaks decrease when v is doubled (dashed lines). The values of $\eta_i$ and $Bi_i$ for both cases are listed in the legend of the figure. Note that $\eta_i \propto 1/v$ and the comparisons are based on the dimensionless time. The effect of increasing v is somewhat similar to that of decreasing $D_{pi}$, since both result in the increase of $Bi_i$ and the decrease of $\eta_i$. But the increasing of $v$ reduces the sharpness and resolution of the peaks more severely because the increase of

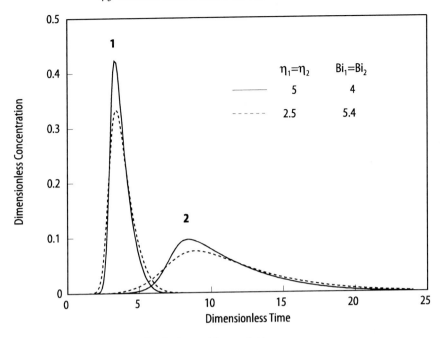

**Fig. 4.7.** Effect of interstitial velocity on binary elution

$Bi_i$ is smaller than that in the case of decreasing $D_{pi}$. Note that in Fig. 4.7, the retention times are not affected by the changes in the $v$ values, since the retention times here are dimensionless.

## 4.3 Effect of Mass Transfer in a Case with Unfavorable Isotherm

In elution chromatography, a peak's front is often sharper than its rear boundary if the isotherm is in the nonlinear range and the isotherm is of favorable type, i.e., concave downward or $\partial^2 C_p^* / \partial C_p^2 < 0$, of which the Langmuir isotherm is a typical example. This is due to the well-known self-sharpening effect of favorable isotherms [47, 72].

Some adsorption systems, namely cooperative adsorption systems, have unfavorable isotherms. It was found that in elution, when the isotherm is of unfavorable type, a peak's front tends to get diffused and its rear boundary sharpened [47, 72]. Such a phenomenon has also been observed in experiments and is well-known in nonlinear chromatography. This is generally true for systems with fast mass transfer rates. For systems with slow mass transfer rates this may not be the case. Figure 4.8 shows that a single

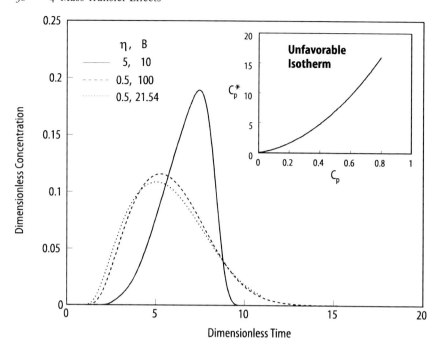

**Fig. 4.8.** Effect of mass transfer on an elution system with an unfavorable isotherm

component elution with an inert mobile phase gives a peak of anti-Langmuirian asymmetry. When $k_i$ or $D_{pi}$ is decreased to some extent the peak symmetry is reversed to that of Langmuir type. In Fig. 4.8, a parabolically shaped unfavorable isotherm is used for the simulation. This phenomenon of peak shape reversal due to mass transfer effects may be attributed to the fact that the diffusive effect of slow mass transfer affects the peak front more severely than its rear boundary. An experimental proof should be helpful.

# 5 Interference Effects in Multicomponent Chromatography

## 5.1 Introduction

Analytical chromatography usually involves small and dilute samples. Thus, interference effects among different sample components are often negligible. With the rapid growth of biotechnology, preparative- and large-scale chromatography become more and more important. High feed concentrations and large sample volumes are often used to increase productivity. In such cases, interference effects may no longer be ignored.

Systematic studies of interference effects in the literature are based mostly on the equilibrium theory [7, 8, 24, 73]. They assume a direct local equilibrium between the liquid phase and the stationary phase, and neglect mass transfer effects. In this chapter, the general multicomponent rate model described in Chap. 3 is used to study the multicomponent interference effects. The model is able to describe some important phenomena such as roll-up in all the three major modes of chromatography – frontal, displacement and elution – under mass transfer conditions. The use of the rate model that considers various mass transfer mechanisms gives a more accurate account, and thus helps the visualization of the dynamics of the preparative- and large-scale chromatographic processes.

Tiselius [74] was the first to use the phrase "displacement effect" to describe the competition for binding sites in multicomponent elution. It will be shown in this chapter that the displacement effect is, in fact, the dominating factor in multicomponent interactions that are directly attributed to the competition for binding sites among different components, and this effect exists in all the three major operational modes. Many observed multicomponent interactions due to competitive adsorption can be satisfactorily explained using this simple concept. Although a few systems exist with synergistic (cooperative) isotherms [7] where the presence of other solutes enhances adsorption, the competitive isotherms are the most common type in practical operations [75].

## 5.2 Computer Simulation and Discussion

The general rate model presented in Chap. 3 with the multicomponent Langmuir isotherm is used to study the interference effects here. The conclusions in most cases can be readily extended to multicomponent systems with other types of competitive isotherms. For comparison and simplicity, the component mixing effect on some physical properties, such as diffusion and mass transfer coefficients, is ignored. All the computer simulations have been carried out on a SUN 4/280 computer with the Fortran 77 code RATE.F. Parameter values used for simulations are listed in Table 5.1, or mentioned during discussions. In all cases $\varepsilon_b=0.4$ and $\varepsilon_p=0.5$. For elution cases, sample sizes are $\tau_{imp}=0.1$, or mentioned otherwise.

### 5.2.1 Displacement Mode

The displacement effect is most noticeable and also relatively well-known in displacement chromatography. Figure 5.1 (solid lines) shows a simulated chromatogram (effluent history) of a stepwise displacement process in which component 2 (displacer) is introduced at $\tau=0$ to a column pre-saturated with

**Table 5.1.** Parameter values used for simulation in Chap. 5[a]

| Figure(s) | Species | Physical Parameters | | | | | Numerical Parameters | |
|---|---|---|---|---|---|---|---|---|
| | | $Pe_{Li}$ | $\eta_i$ | $Bi_i$ | $a_i$ | $b_i \times C_{0i}$ | Ne | N |
| 5.1 | 1 | 300 | 10 | 6 | 3 | $6 \times 0.1$ | 12 | 2 |
| | 2 | 300 | 15 | 8 | 2 | $4 \times 0.4$ | | |
| | 1 | 200 | 10 | 10 | 2 | $2 \times 0.2$ | 25 | 1 |
| 5.2 | 2 | 200 | 10 | 10 | 30 | $30 \times 0.2$ | | |
| | 3 | 200 | 10 | 10 | 80 | $80 \times 0.2$ | | |
| | 1 | 300 | 1 | 20 | 1 | $2 \times 0.1$ | | |
| 5.3, 5.4 | 2 | 300 | 1 | 20 | 10 | $20 \times 0.1$ | 8 | 2 |
| | 3 | 300 | 1 | 20 | 20 | $40 \times 0.1$ | | |
| 5.5 | 1 | 300 | 1 | 20 | 1 | $20 \times 0.1$ | 8 | 2 |
| | 2 | 300 | 1 | 20 | 10 | $200 \times 0.1$ | | |
| | 1 | 300 | 30 | 8 | 1 | $10 \times 1$ | 18 | 1 |
| 5.6 | 2 | 400 | 40 | 7 | 4 | $40 \times 0.1$ | | |
| | 3 | 500 | 90 | 6 | 9 | $90 \times 0.1$ | | |
| 5.7 | 1 | 300 | 40 | 10 | 0.4 | $0.8 \times 0.1$ | 12 | 1 |
| | 2 | 350 | 50 | 9 | 4 | $8 \times 0.1$ | | |
| 5.8 | 1 | 300 | 40 | 10 | 2 | $4 \times 0.1$ | 11 | 1 |
| | 2 | 350 | 50 | 9 | 4 | $8 \times 0.1$ | | |
| | 1 | 300 | 30 | 8 | 4 | $10 \times 0.1$ | 20 | 1 |
| 5.9 | 2 | 400 | 40 | 7 | 16 | $40 \times 0.1$ | | |
| | 3 | 500 | 90 | 6 | 36 | $90 \times 0.1$ | | |
| | 1 | 300 | 40 | 10 | 0.4 | $0.8 \times 0.1$ | 14 | 1 |
| 5.11 | 2 | 350 | 50 | 9 | 4 | $8 \times 0.1$ | | |
| | 3 | 350 | 50 | 9 | 10 | $20 \times 0.1$ | | |

component 1 via a step change. A roll-up peak appears in the concentration profile of component 1, which is a clear indication of the displacement effect. The concentration profile of component 1 is sharpened compared with the dashed line that represents the corresponding desorption operation when only an inert mobile phase is used to "wash out" component 1 from the column. In other words, the use of the displacer reduces the tailing and thus concentrates component 1. This is also evident in Fig. 5.2 that shows a simulated chromatogram of a binary displacement system, in which components 1 and 2 are introduced to the column via a frontal adsorption lasting $\tau_{imp}=4.0$ before component 3 (displacer) is pumped into the column. In this volume overload case, component 2 has two peaks between which the roll-up peak is due to the displacement effect from the displacer (component 3). Such a concentrating effect was proven by experiments carried out by Helfferich [76]. In Fig. 5.2 there is also a roll-up peak for component 1 that is the result of the displacement effect from component 2. The first smaller component 2 peak should not be mistaken as a displacement band of a separate component.

Competition can be viewed as a mutual interaction. The displaced component, while being displaced, in return also exerts some influence on the displacer. Figure 5.1 illustrated that the concentration front of the displacer is

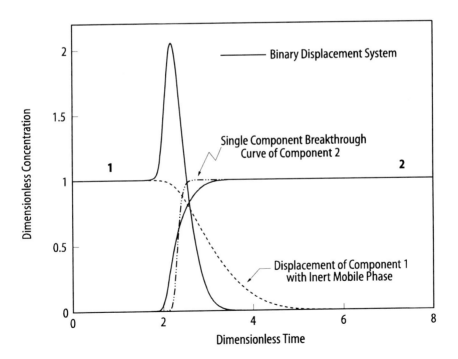

**Fig.5.1.** Two-component stepwise displacement process

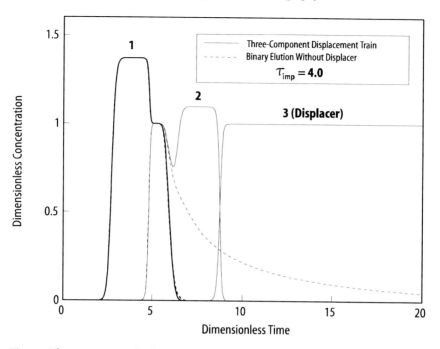

**Fig. 5.2.** Three-component displacement system

diffused by component 1 in the displacement process. The concentration profile of component 2 is actually its breakthrough curve under the interference of component 1. As compared with the breakthrough curve of pure component 2 (double-dotted line in Fig. 5.1), the concentration front of component 2 becomes diffused due to component 1.

The concentrating effect and roll-up phenomenon in displacement chromatography with negligible mass transfer effects have been predicted by the ideal theories including the interference theory [5-7, 47, 77]. The general model presented here describes the roll-up phenomenon under mass transfer conditions.

### 5.2.2 Frontal Adsorption Mode

Figure 5.3 shows a simulated chromatogram of a binary frontal adsorption process, in which component 1 has a weaker affinity than component 2. The concentration profile of component 1 reaches a maximum that is larger than its feed concentration before leveling off. This roll-up phenomenon is the result of a displacement effect. The concentration front of component 1, which has a weaker affinity, migrates faster than the concentration front of component 2 inside the column. Component 1 takes advantage of the relative

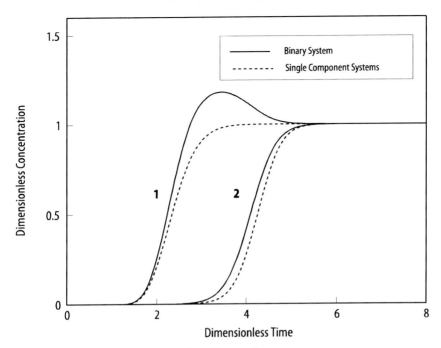

**Fig. 5.3.** Binary frontal adsorption with a roll-up peak

absence of component 2 and initially occupies a disproportionate share of binding sites. When the concentration front of component 2 catches up, it displaces some portion of component 1 such that the concentration of component 1 may exceed its feed value causing the roll-up. The column finally reaches adsorption equilibrium, and each component occupies its share of binding sites according to the governing multicomponent isotherms. Experimental observations and simulations for the roll-up in frontal adsorption with mass transfer effects have been reported by many researchers [24, 28, 29, 41, 78-83].

A comparison of the breakthrough curves of the binary system and their corresponding pure component breakthrough curves in Fig. 5.3 indicates that earlier breakthroughs result for both components in the binary system. This reflects that the dimensionless hold-up capacity of each component in the column is lower compared to the corresponding pure component case.

Figure 5.4 shows a ternary system in which a third component, which has a stronger affinity than the other two, is added to the binary system shown in Fig. 5.3. Two roll-up peaks appear and, by the same token, they can be explained by the displacement effect. The last component, which has the strongest affinity, does not roll-up in any isothermal frontal adsorption case. It is worthwhile to note that the component in the middle (component 2) displaces component 1, while it is displaced by component 3 itself.

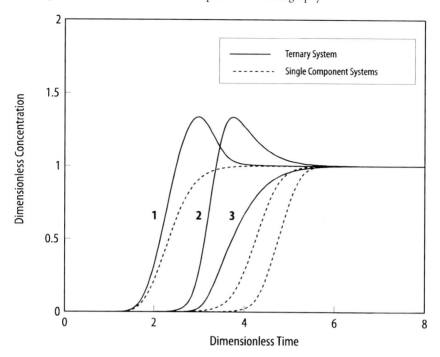

**Fig. 5.4.** Ternary frontal adsorption with two roll-up peaks

Roll-up peaks do not always exist or are noticeable in frontal adsorptions, especially when the saturation capacities of the components are low, or the components have very similar physical properties. Figure 5.5 has the same conditions as Fig. 5.3, except that its $b_i$ values are ten times those for Fig. 5.3, thus, the saturation capacity ($C_i^\infty = a_i / b_i$) values for Fig. 5.5 are 1/10 of those for Fig. 5.3. In Fig. 5.5, the roll-up phenomenon is not noticeable, but the displacement effect is still evident.

### 5.2.3  Elution Mode

Multicomponent elution with an inert mobile phase results in a shortened retention time for each component (Fig. 5.6). The retention times here are based on the first moments rather than the positions of peak maxima. The peak height of component 1, which has a weaker affinity than component 2, is increased, indicating that less band spreading occurs as compared with the corresponding single component case. For the component 2 peak, its front is diffused and tail reduced. Conversely, the peak height of component 3 is significantly decreased and its front is severely diffused.

These observations again can be explained by a displacement effect. When the three components are migrating inside the column with different speeds

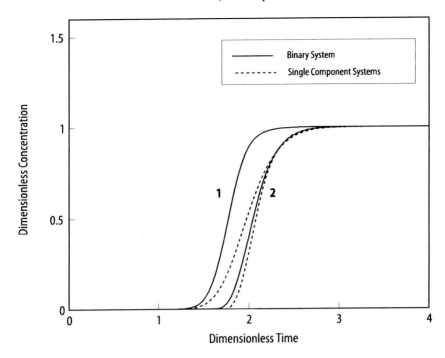

**Fig. 5.5.** Binary frontal adsorption with no roll-up peak

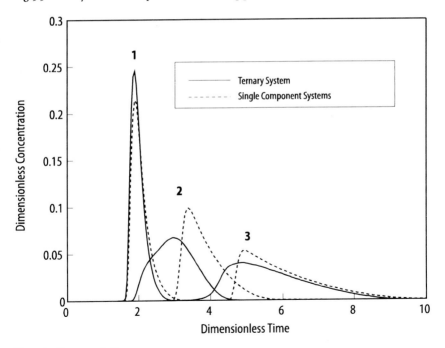

**Fig. 5.6.** Ternary elution

depending primarily on their adsorption affinities, they separate from each other. Since component 2 has a higher affinity than component 1, it travels behind and displaces and concentrates component 1, thus reducing the tail of the component 1 peak. This results in a slightly shorter retention time, a larger peak height, and less band spreading for component 1. The displacement effect in such a case has been mentioned by other researchers [47, 74].

Mutual displacement causes the portion of component 2 that is in the mixing zone with component 1 to migrate faster than for the single component case inside the column, while the unmixed portion of component 2 migrates with the same speed as the single component case. This causes the diffusion of the front of the component 2 peak. In comparison, the displacement effect of component 3 on component 2 reduces the tail of the component 2 peak. In return, component 2 diffuses the front of the component 3 peak. Since component 3 elutes last, the tail's end-point of its peak hardly changes as compared to the single component case. The effect of surrounding components is further illustrated by component 2 in Fig. 5.6 where the diffusion effect of component 1 reduces the peak height of component 2, while the displacement effect of component 3 tends to do the opposite. Therefore, the net effect of these two influences will determine the relative peak height of component 2.

The influence of the displacement effect on nonlinear multicomponent elution is summarized in Table 5.2, with the understanding that the effects listed in the table may not always be noticeable depending on the severity of the displacement effect. The severity of the displacement effect depends on the level of competition among all the components and the nonlinearity of the system. In multicomponent elution five factors have impacts on the displacement effect.

**Table 5.2.** Summary of multicomponent elution (compared with single component elutions)

| Peak Position in Chromatogram | Retention Time (First Moment) | Peak Height | Front Flank | Tailing |
|---|---|---|---|---|
| First peak | decreases | increases | sharpens | decreases |
| Middle Peak(s) | decreases | increases or decreases | diffuses | decreases |
| Last Peak | decreases | decreases | diffuses | a |

a The tail end-point does not change much while the tail may become flatter (see Figs. 5.6 and 5.8)

### 5.2.3.1  Adsorption Equilibrium Constants

As the values of $b_i$ increase, the nonlinearity of the isotherm and the competition for binding sites also increase. This escalates the displacement effect. If the values of $b_i$ in a binary elution system are similar, the contact time between the two components is maximized as they migrate through the column and separate from each other. This increases the displacement effect. Figure 5.7 has the same conditions as Fig. 5.8, except that in Fig. 5.8, the affinity of component 1 is larger, thus closer to that of component 2 (see Table 5.1). Compared with Fig. 5.7, the displacement effect in Fig. 5.8 is obviously more pronounced.

### 5.2.3.2  Low Adsorption Saturation Capacity

A lower saturation capacity means fewer binding sites, and often increased competition for binding sites, especially in a system with large $b_i$ values. In Fig. 5.6, a system with a small saturation capacity ($C_i^\infty$) was used in order to

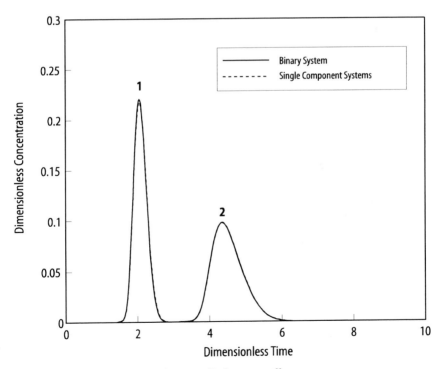

**Fig. 5.7.** Binary elution showing almost no displacement effect

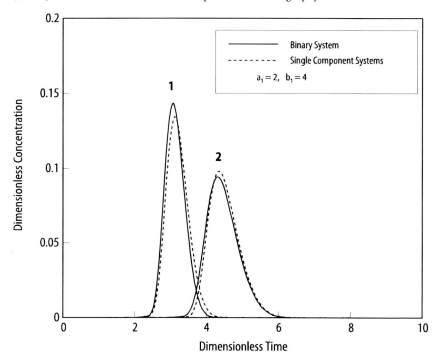

**Fig. 5.8.** Binary elution with increased $a_1$ and $b_1$ values

show a case with pronounced displacement effects. Figure 5.9 is obtained from Fig. 5.6 by increasing the values of $a_i$ for the three components by four-fold. The displacement effect is more noticeable in Fig. 5.6 than in Fig. 5.9.

### 5.2.3.3  High Sample Feed Concentration (Concentration Overload)

Increasing $C_{0i}$ is equivalent to increasing $b_i$ and reducing $C_i^\infty$ proportionally, as is shown by the isotherm expression, Eq. (3-17). Thus, the displacement effect escalates when the feed concentrations of the sample are increased.

### 5.2.3.4  Large Sample Size (Volume Overload)

When a large sample size is used, the contact time between the components will increase, thus making the displacement effect more noticeable. Figures 5.7 and 5.10 have the same conditions except that in Fig. 5.10, the sample size ($\tau_{imp}$=2.5) is much larger than that in Fig. 5.7 ($\tau_{imp}$=0.1). The first half of the effluent history in Fig. 5.10 actually represents the concentration profiles of

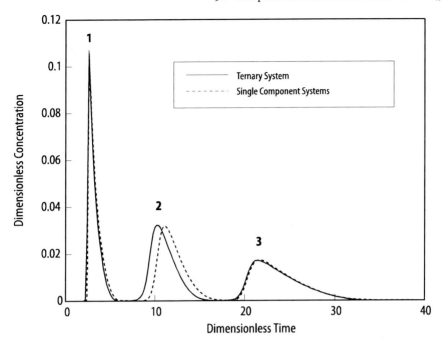

**Fig. 5.9.** Ternary elution with increased saturation capacity

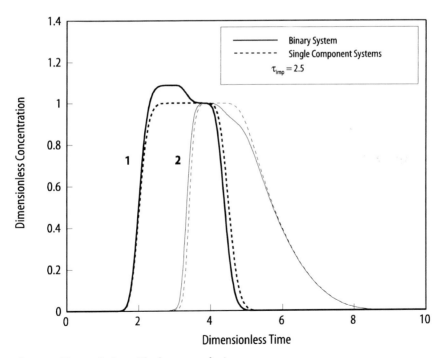

**Fig. 5.10.** Binary elution with a large sample size

the frontal adsorption curves with a roll-up peak, due to severe volume overload.

The use of large sample sizes in elution is not rare. In order to promote
column throughput, the column is often overloaded in terms of either sample
size or sample concentration [18, 49, 72, 84, 85]. Overload generally increases
the nonlinearity of the system and thus the displacement effect.

### 5.2.3.5 More Component(s)

Adding more component(s) in the sample will increase the competition for
binding sites among components. It also increases the nonlinearity of the
isotherms, thus escalating the displacement effect. The increased displacement effect in Fig. 5.11 is obtained by adding one more component to the case
presented in Fig. 5.7.

When an additional component is present as a competing modifier in the
mobile phase, the displacement effect becomes rather complicated. The peaks
corresponding to the concentration profile of the modifier in a chromatogram are often referred to as system peaks, which will be discussed in the
Chap. 6.

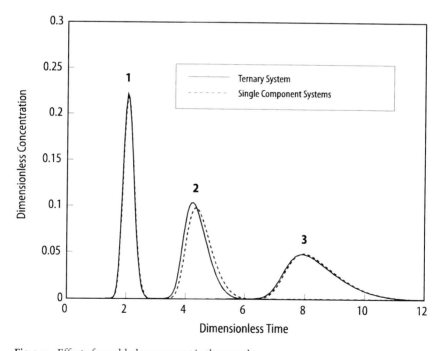

**Fig. 5.11.** Effect of an added component in the sample

## 5.3 Summary

For multicomponent chromatography involving competitive isotherms, the dominating interference effect can be attributed to a displacement effect, which occurs not only in the displacement mode but also in the other two major modes of chromatography, frontal and elution. Five factors that may escalate the displacement effect in elution chromatography have been investigated. In short, these five factors either promote competition for binding sites among components or prolong such competition, or both. From a mathematical point of view, these factors can be interpreted as being able either to increase or to prolong the nonlinearity of the isotherms. It has also been shown that roll-up exists not only in frontal adsorption and displacement, but also in elution. It has been demonstrated that the displacement effect tends to reduce peak tails of the displaced components, while the concentration front of the displacer's peak is diffused by the displaced components in all the three major modes of chromatography.

The use of a general nonlinear multicomponent rate model provides a systematic study of interference effects in multicomponent chromatography. The model accounts for various diffusional and mass transfer effects. The graphical representation of the results aids the visualization of multicomponent interactions, and thereby promotes a better understanding of the primary causes of the interference effects. The discussion presented here may also be useful in the optimization of chromatographic separation processes.

# 6 System Peaks in Multicomponent Elution

## 6.1 Introduction

In isocratic elution chromatography, a modifier is often added to the mobile phase in order to compete with sample solutes for binding sites [2]. This helps reduce the retention time and band spreading of the sample solutes.

Peaks attributed to the modifier in an elution chromatogram are called system peaks [86-88]. A positive system peak, which is above the baseline value of the modifier concentration, is called a displacement peak [20, 89]. A negative one, which is below the baseline value, is called a vacancy peak [89]. Solms et al. [20] used a plate model to simulate three cases of single component elution with a mobile phase containing a competing modifier. Another group of researchers [90, 91] simulated binary elutions with a competing modifier using a semi-ideal model with nonlinear multicomponent Langmuir isotherms. They also performed experiments that qualitatively proved some of their model predictions.

Two different types of sample are used for elution chromatography with the mobile phase containing a modifier. The first type, named Type I sample in this chapter, consists of those samples that are prepared by dissolving sample solutes in a solution that has the same composition as the mobile phase, thus the feed stream to the column contains the competing modifier with a constant concentration. This kind of system is a strictly isocratic elution process if the modifier concentration in the feed is constant. Modeling of system peaks with Type I samples was first carried out by Solms et al. [20]. The second type of sample, Type II sample, consists of those samples that are prepared based on an inert (blank) solution, i.e., the samples contain no modifier. In such cases, system peaks have different patterns from those with Type I samples because of the deficit of modifier introduced during the sample injection. Experiments with both types of sample were carried out by Levin and Grushka [88]. They also investigated elution systems containing more than one modifier in the mobile phase.

This chapter extends the previous theoretical studies in the literature on elution chromatography with a competing modifier, which has a constant concentration in the mobile phase, using the general rate model described in Chap. 3. The effect of modifier on the elution performance of binary-solute systems with a Type I or Type II sample will be studied and system peak patterns will be summarized for both cases. Binary elution with two modifiers in the mobile phase will also be briefly discussed.

## 6.2 Boundary Conditions for the General Rate Model

The modifier is treated as one of the components in the governing equations of the rate model. The multicomponent Langmuir isotherm is used, in which the modifier is considered as one of the competing components.

The following boundary conditions are needed for the modifier. For a modifier in systems with Type I sample,

$$C_{fi}(\tau)/C_{0i} = 1 \quad . \tag{6-1}$$

For a modifier in systems with Type II sample,

$$C_{fi}(\tau)/C_{0i} = \begin{cases} 1 & 0 \leq \tau \leq \tau_{imp} \\ 0 & \text{else} \quad . \end{cases} \tag{6-2}$$

## 6.3 Results and Discussion

The Fortran 77 code RATE.F is used for simulations in this chapter to study system peaks. Parameter values used for simulations are listed in Table 6.1, or mentioned during discussions. In all runs, $\varepsilon_p=\varepsilon_b=0.4$, and the rectangular sample size is $\tau_{imp}=0.1$, unless otherwise specified. Since the competing modifier is considered as a competing component in the multicomponent Langmuir isotherm, a binary elution with a competing modifier in the mobile phase constitutes a threecomponent system.

### 6.3.1 Modifier is weaker than sample solutes

Figure 6.1 (solid lines) shows a simulated effluent history (chromatogram) of a binary elution with Type I sample and a competing modifier (component 3) in the mobile phase. Components 1 and 2 are the two sample solutes. The affinity of the modifier is smaller than those of the sample components. Note that the scale for the modifier concentration shown in Fig. 6.1 (as well as in all other figures) is $(c_{b3}-1)$. The actual baseline value for the modifier concentration is $(c_{b3}=1)$. By transforming the baseline value to zero (i.e., $c_{b3}-1=0$), the effluent history becomes more presentable. A negative system peak does not indicate negative concentrations, but rather concentration values that are below the baseline value.

**Table 6.1.**    Parameter values used for simulation in Chap. 6[*]

| Figure(s) | Species | Physical Parameters | | | | | Numerical Parameters | |
|---|---|---|---|---|---|---|---|---|
| | | $Pe_{Li}$ | $\eta_i$ | $Bi_i$ | $a_i$ | $b_i \times C_{0i}$ | Ne | N |
| 6.1, 6.2 | 1 | 300 | 8 | 20 | 5 | 2.5 ×0.2 | 8 | 2 |
| | 2 | 400 | 9 | 12 | 10 | 5 ×0.2 | | |
| | 3 | 350 | 9.5 | 9 | 2 | 1 ×0.1 | | |
| 6.3, 6.4 | 1 | 300 | 8 | 20 | 5 | 2.5 ×0.2 | 8 | 2 |
| | 2 | 400 | 9 | 12 | 10 | 5 ×0.2 | | |
| | 3 | 350 | 9.5 | 9 | 7 | 3.5 ×1 | | |
| 6.5, 6.6 | 1 | 300 | 8 | 20 | 5 | 2.5 ×0.2 | 7 | 2 |
| | 2 | 400 | 9 | 12 | 10 | 5 ×0.2 | | |
| | 3 | 350 | 9.5 | 9 | 20 | 10 ×0.1 | | |
| 6.7 | 1 | 300 | 8 | 20 | 5 | 2.5 ×0.2 | 8 | 2 |
| | 2 | 400 | 9 | 12 | 10 | 5 ×0.2 | | |
| | 3 | 350 | 9.5 | 9 | 40 | 20 ×0.1 | | |
| 6.8 | 1 | 300 | 8 | 20 | 5 | 2.5 ×0.2 | 8 | 2 |
| | 2 | 400 | 9 | 12 | 10 | 5 ×0.2 | | |
| | 3 | 350 | 9.5 | 9 | 100 | 50 ×0.1 | | |
| 6.11-6.13 | 1 | 300 | 8 | 20 | 5 | 5 ×0.4 | 9 | 2 |
| | 2 | 300 | 8 | 20 | 20 | 20 ×0.4 | | |
| | 3 | 300 | 8 | 20 | 40 | 40 ×0.2 | | |
| 6.15 | 1 | 300 | 8 | 20 | 5 | 2.5 ×0.2 | 8 | 2 |
| | 2 | 400 | 9 | 12 | 6.6 | 3.3 ×0.2 | | |
| | 3 | 350 | 9.5 | 9 | 2 | 1 ×0.1 | | |
| 6.16 | 1 | 300 | 8 | 20 | 1 | 0.5 ×0.2 | 9 | 2 |
| | 2 | 400 | 9 | 12 | 10 | 5 ×0.2 | | |
| | 3 | 350 | 9.5 | 9 | 100 | 50 ×0.1 | | |
| 6.17 | 1 | 300 | 8 | 20 | 5 | 2.5 ×0.2 | 8 | 2 |
| | 2 | 400 | 9 | 12 | 6.6 | 3.3 ×0.2 | | |
| | 3 | 350 | 9.5 | 9 | 7 | 3.5 ×0.48 | | |
| 6.19, 6.20 | 1 | 200 | 6 | 9 | 6 | 3 ×0.25 | 8 | 2 |
| | 2 | 200 | 6 | 9 | 20 | 10 ×0.2 | | |
| | 3 | 200 | 6 | 9 | 2 | 1 ×0.2 | | |
| | 4 | 200 | 6 | 9 | 20 | 10 ×0.2 | | |

The case shown in Fig. 6.1 gives one positive system peak and two negative system peaks that are due to the displacement effect of two sample solutes on the modifier. A mass balance of each species has been checked to evaluate the accuracy of the numerical solution. For the modifier, the numerical integration (using the Fortran 77 code AREA.F) of the concentration profile of the modifier ($c_{b_3}-1.0$) in Fig. 6.1, which consists of 400 data points, from $\tau=1$ to $\tau=15$ gives a value of 0.0000, which is in agreement with its theoretical value zero. For the sample solutes, mass balances are also held.

Figure 6.1 (dashed lines) also shows the binary elution case in the absence of the modifier. It is evident that the use of a modifier results in the decrease of the retention time and the spreading of the band and the increase of peak height of each sample solute. Figure 6.2 shows an effluent history with a Type

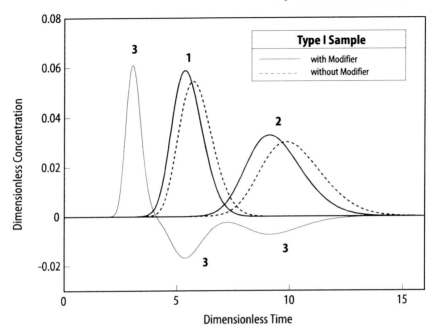

**Fig. 6.1.**  Binary elution with a weak modifier (Type I sample)

II sample. Other conditions for Fig. 6.2 are the same as in Fig. 6.1. It can be seen that in Fig. 6.2, there are three negative system peaks and no positive ones. The numerical integration of the concentration profile for the modifier (component 3) of the three system peaks has been found to be −0.1000. This negative value indicates the deficit of modifier introduced during sample injection. The deficit quantity is equivalent to the sample size, $\tau_{imp}$=0.1. In Fig. 6.2 the peak at the front is a negative system peak, instead of a positive one shown in Fig. 6.1, because the large negative system peak induced by the deficit of the modifier during sample injection negates the positive system peak. This can easily be verified by examining the concentration profile of the modifier when a blank sample, which contains only an inert carrier liquid, is employed. This is shown in Fig. 6.2 (dashed line). It gives only a single large negative peak, and the peak area is found to be equal to the injection pulse size, $\tau_{imp}$=0.1 by numerical integration.

Positive system peaks do occur involving a Type II sample, if the positive system peak overcomes the negative one due to sample introduction, as shown shortly. The number and direction (positive/negative, i.e., upward/downward) of system peaks for the modifier are determined primarily by the sample type and their relative affinity to those of the sample solutes, and of course, the number of sample solutes.

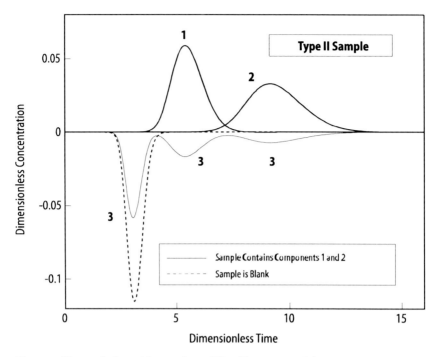

**Fig. 6.2.** Binary elution with a weak modifier (Type II sample)

### 6.3.2 Modifier affinity is between those of sample solutes

Figure 6.3 gives the effluent history for the case shown in Fig. 6.1 except that the affinity of the modifier is between those of the two sample solutes. Figure 6.3 shows one positive system peak and two negative ones, which are similar to those in Fig. 6.1. However, the retention time of the positive system peak is prolonged and the peak is sharpened. Both changes are due to the increase of affinity of the modifier. Because of the increase, there are more modifier molecules adsorbed onto the stationary phase that can be dislodged by the sample solutes. If the modifier has no affinity to the column packing, its concentration profile will be flat. On the other hand, if the affinity of the modifier further increases when its affinity is already not far from the leveling off range of the Langmuir isotherm, the increase of modifier's loading in the stationary phase can be overshadowed by the affinity increase that could make it too difficult to be dislodged by the sample solutes. In such cases, the increase of modifier affinity may result in a reduced positive system peak at the front.

Figure 6.4 has the same conditions as Fig. 6.3, except that a Type II sample is used in Fig. 6.4. The effluent history shown in Fig. 6.4 gives one positive and two negative system peaks, quite different from the case shown in Fig. 6.2,

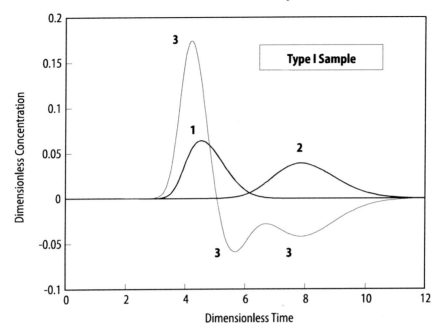

**Fig. 6.3.**  Modifier affinity is between those of sample solutes (Type I sample)

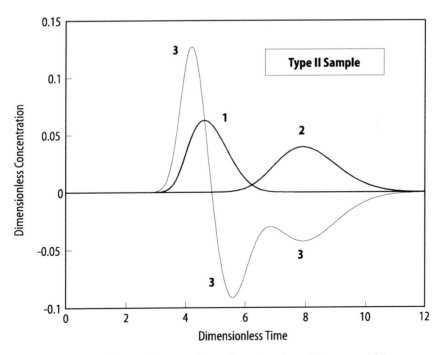

**Fig. 6.4.**  Modifier affinity is between those of sample solutes (Type II sample)

which has the same conditions as Fig. 6.4, except that Fig. 6.4 has a stronger modifier. This is because the displacement effects from components 1 and 2 cause a larger positive system peak, and it overcomes the negative system peak that is caused by the deficit of modifier in the sample. That the positive system peak in Fig. 6.4 is smaller than the one in Fig. 6.3 is in agreement with this argument.

### 6.3.3  Modifier is stronger than sample solutes

Figure 6.5 shows a case in which the affinity of the modifier is stronger than both sample solutes. There are two positive system peaks and one negative one in the figure. The first positive system peak partially overlaps with the component 1 peak, and it departs from the component 1 peak when the component 2 peak starts to take off. The corresponding case with a Type II sample is shown in Fig. 6.6 that gives a similar system peak pattern.

In Fig. 6.5, if the affinity of the modifier is further increased, the first positive system peak will no longer overlap the component 1 peak, as is shown clearly in Fig. 6.7. Figure 6.8 shows a degenerated case that is obtained by simply increasing the affinity of the modifier shown in Fig. 6.7. It is obvious that the merger of two positive system peaks in Fig. 6.8 is due to the partial overlapping of the component 1 peak with the component 2 peak.

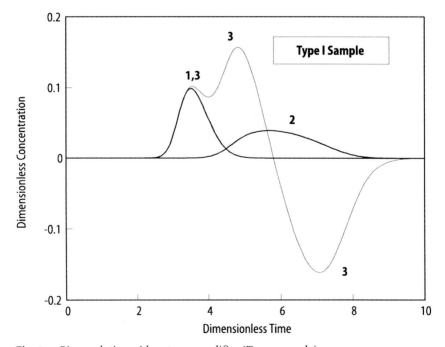

**Fig. 6.5.** Binary elution with a strong modifier (Type I sample)

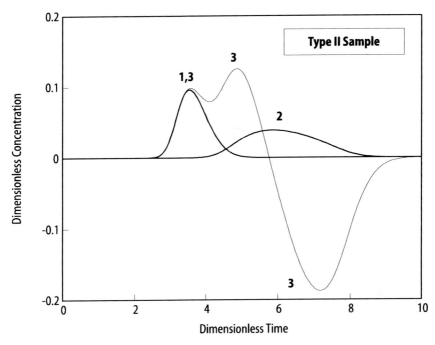

**Fig. 6.6.** Binary elution with a strong modifier (Type II sample)

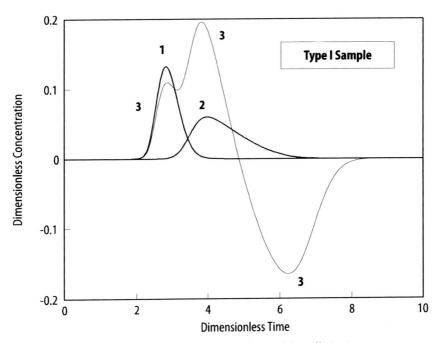

**Fig. 6.7.** Same conditions as Fig. 6.5, except that the modifier affinity is stronger

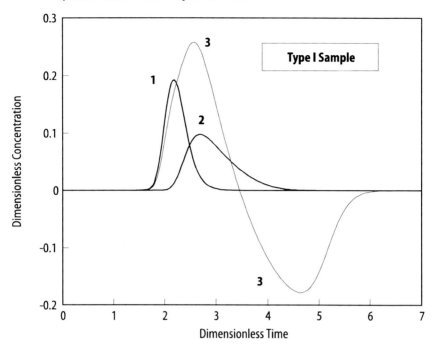

**Fig. 6.8.** Same conditions as Fig. 6.7, except that the modifier affinity is stronger

### 6.3.4  Effect of modifier concentration on system peak patterns

The increase of modifier concentration obviously reduces retention times of the sample solutes because the modifier competes with sample solutes for binding sites. It also affects system peaks. Figure 6.9 shows a case in which the modifier concentration is ten times higher than that in Fig. 6.3. Comparing Fig. 6.3 with Fig. 6.9, it can be seen that the system peaks in Fig. 6.9 are much smaller than those in Fig. 6.3. This means that the disturbance caused by the sample solutes to the concentration profile of the modifier becomes smaller if the concentration of the modifier increases. It also implies, that if the modifier concentration is sufficiently large, its concentration in the system can be considered as a constant. This simplifies the simulation. Note that in all figures the concentration scale is dimensionless. Thus, a smaller peak does not necessarily mean a smaller dimensional concentration.

Figure 6.10 shows a case with a Type II sample, in which the modifier concentration is ten times higher than in Fig. 6.4. The increase of modifier concentration changes the first system peak from a positive one (Fig. 6.4) to a negative one (Fig. 6.10). The reversal of the peak direction occurs because when the modifier concentration is increased, the negative system peak,

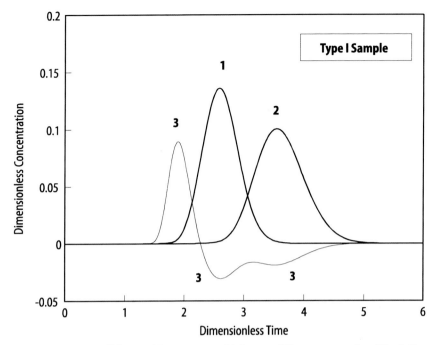

**Fig. 6.9.** Same conditions as Fig. 6.3, except higher modifier concentration ($C_{o_3}$=1.0)

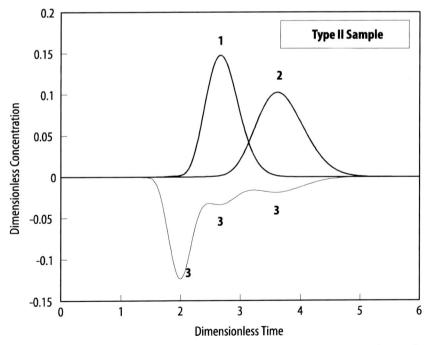

**Fig. 6.10.** Same conditions as Fig. 6.4, except higher modifier concentration ($C_{o_3}$=1.0)

which is caused by the deficit of the modifier during sample injection, overcomes the positive system peak caused by the displacement effect from the sample solutes on the modifier.

### 6.3.5  Effect of modifier on sample solutes

The retention time and resolution of the two solute peaks are both unnecessarily high in Fig. 6.11 for a complete separation of the two components. Adding a proper modifier may reduce the process duration while still achieving a baseline separation. Figure 6.12 shows the effect of an added modifier. The baseline separation of the two sample solutes is achieved while the elution duration is cut by four-fifths. The concentrations of the peaks are much higher and the band spreading of these peaks is largely reduced when the modifier is used. This is because of the displacement effect from the modifier. Figure 6.13 has the same conditions as Fig. 6.12, except that in Fig. 6.13 a Type I sample is employed. The result shown in Fig. 6.13 is similarly desirable.

Figures 6.12 and 6.13 show that at low modifier concentration levels, the phenomenon of peak shape reversal also occurs if the adsorption equilibrium constant of the modifier ($b_3$) is high enough and the adsorption capacity is

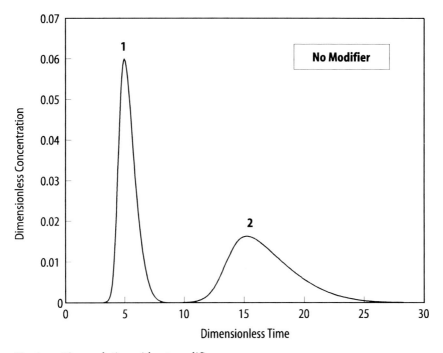

Fig. 6.11. Binary elution without modifier

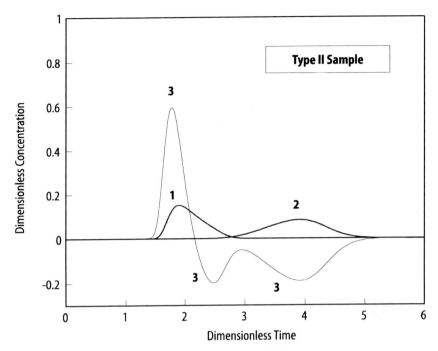

**Fig. 6.12.**  Effect of an added modifier (Type II sample)

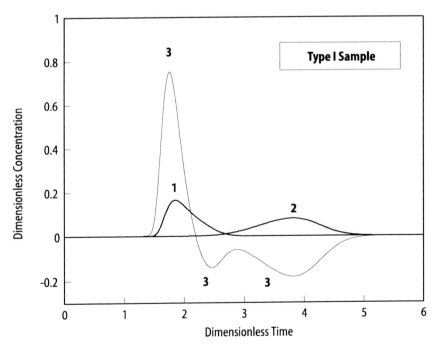

**Fig. 6.13.**  Effect of an added modifier (Type I sample)

low. Interestingly, in the two figures, component 1 peak still retains its Langmuir type peak shape while the peak shape of component 2 becomes anti-Langmuirian type that has a smaller tail than its front flank.

### 6.3.6 Effect of sample type

The difference in the system peak pattern due to sample type is quite obvious. Sample type may affect the direction, size and location of system peaks. These effects have been shown during the previous discussion. On the other hand, the sample type also affects the elution pattern of sample solutes. By comparing some of the figures shown in this chapter, one may quickly find that the influence of sample type on sample solutes is usually quite small. This situation may be changed if the sample size is large. Figures 6.3 and 6.4 have the same conditions except the type of sample. The sample size in both cases is $\tau_{imp}=0.1$. Their corresponding cases with $\tau_{imp}=1.0$ are shown in Fig. 6.14. It is clear that the difference in the concentration profiles of the two sample solutes is not small when a large sample size is used.

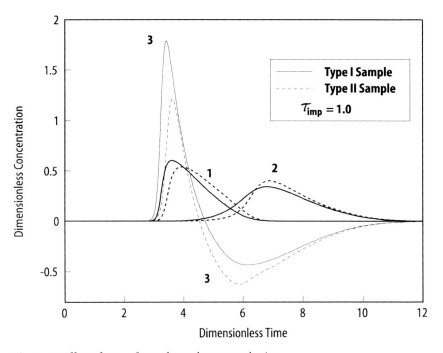

**Fig. 6.14.** Effect of type of sample at a large sample size

### 6.3.7  Effect of sample solutes on the modifier

In the above discussion, it has been pointed out that system peaks are the result of the displacement effects of the sample solutes on the modifier arising from competition for binding sites. This was also revealed by other researchers [20, 90]. If a Type II sample is used, the deficit of modifier during the sample introduction also plays a role that may cause a negative system peak at the front (Fig. 6.2) or reduce the size of the positive system peak at the front (Fig. 6.14). It may even negate the positive system peak (Fig. 6.15).

The relative affinities of sample solutes also affect system peaks as shown in Fig. 6.15. Figure 6.15 has the same conditions as Fig. 6.2, except that in Fig. 6.15 the affinity of component 2 is smaller, thus closer to that of component 1. Figure 6.15 (with a Type II sample) shows that when component 1 and component 2 peaks overlap to some degree, the two corresponding negative system peaks will degenerate into a single one. The comparison of Fig. 6.8 with Fig. 6.16 proves that a partial overlapping of the peaks for sample solutes may also cause the merger of positive system peaks.

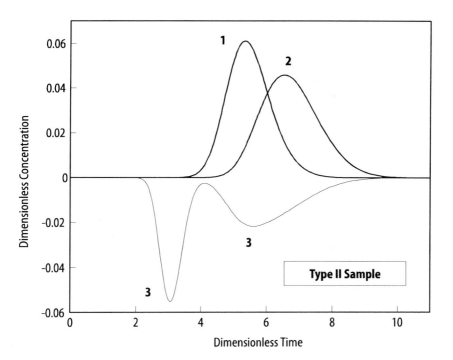

**Fig. 6.15.**  Same conditions as Fig. 6.2, except that Component 2 has a weaker affinity

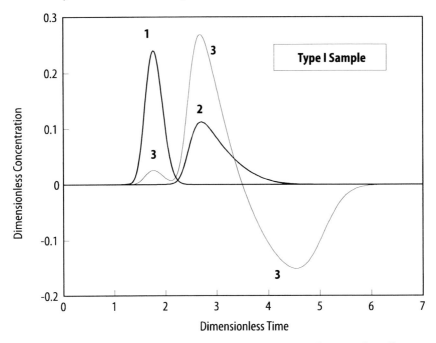

**Fig. 6.16.**  Same conditions as Fig. 6.8, except that Component 1 has a weaker affinity

### 6.3.8  Summary of system peak patterns

Table 6.2 summarizes all possible combinations of system peak patterns for binary elutions with a competing modifier. There are twice as many combinations for cases with a Type II sample as for those with a Type I sample. This table also gives indications for system peak combinations in a single component elution since degenerated cases are included in the table. It is interesting to point out that Fig. 6.17 gives a severely degenerated case in which the overlapping of components 1 and 2 peaks causes the degeneration of their corresponding negative system peaks. The positive displacement peak and the peak that is due to the deficit of modifier during the sample introduction negated each other. Figure 6.18 has the same conditions as Fig.

**Table 6.2.**  Possible system peak combinations in binary elutions

| Sample | System peak combinations (positive peak(s)/negative peak(s)) | | | | | |
|---|---|---|---|---|---|---|
| | I | II | III | IV | V | VI |
| Type I | 1/2 | 1/1 | 2/1 | | | |
| | (Fig. 6.1) | (Fig. 6.8) | (Fig. 6.5) | | | |
| Type II | 0/3 | 0/2 | 0/1 | 1/1 | 2/1 | 1/2 |
| | (Fig. 6.2) | (Fig. 6.15) | (Fig. 6.17) | (Fig. 6.18) | (Fig. 6.6) | (Fig. 6.4) |

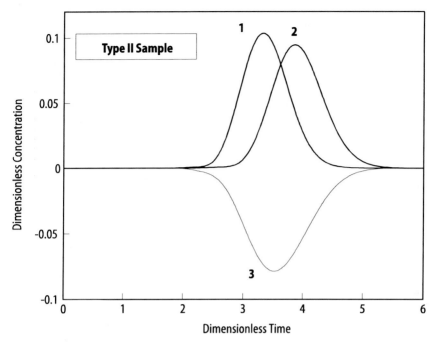

**Fig. 6.17.** Binary elution showing only one system peak (Type II sample)

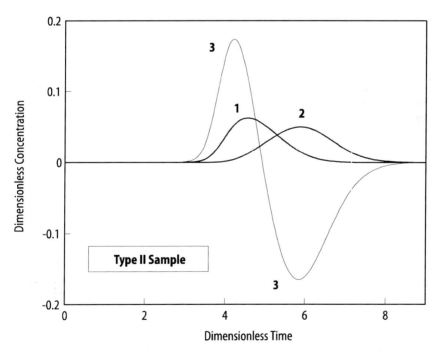

**Fig. 6.18.** Binary elution showing one positive, and one negative system peak, respectively

6.17, except that the modifier concentration is 0.1, which is lower than that in Fig. 6.17. Because of the decrease of the modifier concentration the previously degenerated peak (in Fig. 6.17) becomes very prominent in Fig. 6.18.

In general, both sample types, I and II, can have a maximum of only three system peaks for binary elutions with one competing modifier. For binary elutions with a Type I sample, the minimum number of system peaks should be two because the existence of a positive system peak necessitates a negative system peak in order to meet the mass balance, which requires that the sum of peak areas of positive system peaks be equal to the sum of peak areas of negative system peaks. On the other hand, this requirement does not apply to cases with a Type II sample. In such cases, the minimum number of system peaks is one as shown in Fig. 6.17.

### 6.3.9    Binary elution with two different modifiers

As the discussion above indicates, system peak behavior can be very complex and elusive. The situation can be further complicated if there is more than one modifier in the mobile phase. In practice, multiple modifier cases are not rare. Experiments by Levin and Grushka [88] showed that each modifier gave a different set of system peaks.

Figure 6.19 shows a case involving two sample solutes (components 1 and 2) and two different modifiers (components 3 and 4). The first modifier (component 3) has a weaker affinity than the second modifier (component 4). Figure 6.20 has the same conditions as Fig. 6.19, except that a Type I sample is employed. There is a positive system peak at the tail of the concentration profile of the first modifier (component 3) in both figures. This kind of tail has never been observed in simulations for single modifier systems. Its presence is likely due to the involvement of a second modifier in the system.

## 6.4    Concluding Remarks

The interrelationship between sample solutes and the modifier(s) in elution chromatography has been investigated through computer simulation using a general rate model. It has been concluded that, for binary elutions with one competing modifier in the mobile phase, there are three system peak patterns if Type I samples are used, and six if Type II samples are used. In addition, a binary elution system with two different competing modifiers has been briefly discussed.

This study shows that system peaks can be very complex and may not be fully explained by qualitative arguments, although the ultimate cause behind system peaks may be simply attributed to the displacement effect due to the competitive nature of the isotherms involving all the components in the

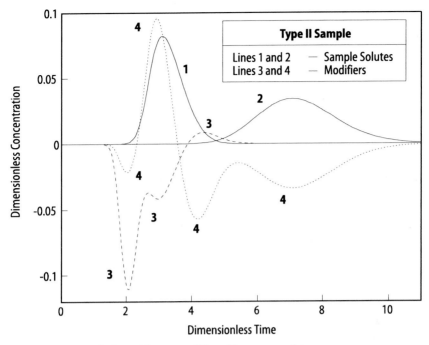

**Fig. 6.19.** Binary elution with two modifiers (Type II sample)

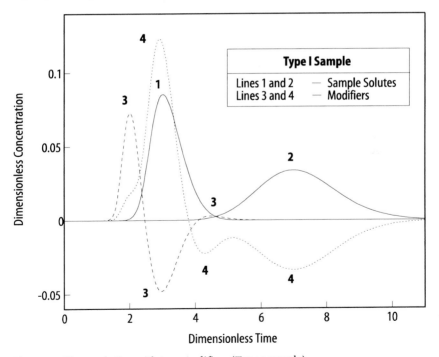

**Fig. 6.20.** Binary elution with two modifiers (Type I sample)

system including the modifier, and the deficit of modifier during a sample introduction if a Type II sample is used.

In gradient elutions, the modifier concentration is continuously changed during the elution process. The situation in gradient elutions is more complicated since the modifiers used in gradient elutions cannot usually be considered as competing with sample solutes for binding sites. They can actually change the binding affinities of the sample solutes with the stationary phase, which is a much more efficient mechanism for the retention time control. The mutual interaction is quite complex. A study of multicomponent nonlinear gradient chromatography using experimental solute-modifier relationships is presented in Chap. 9.

# 7 Multicomponent Adsorption with Uneven Saturation Capacities

## 7.1 Introduction

In chromatographic separations of large biomolecules, such as proteins, using porous adsorbents, a size exclusion effect may be significant. Some large molecules cannot access either part of the small macropores in the adsorbent particles or all the macropores. This is especially possible in chromatographic separations of large proteins. For a multicomponent system involving components with very different molecular sizes, the extent of size exclusion is not the same for all the components. This causes uneven adsorption saturation capacities (based on moles) for the components. The least excluded component tends to have the highest saturation capacity and vice versa.

Modeling of size exclusion coupled with adsorption is a relatively new topic. In size exclusion chromatography with globular sample components, the component that has a larger molecular size has a smaller chance to penetrate the macropores of the particles, and thus it has a smaller retention time. A size exclusion medium should have very little adsorption ability, since adsorption is often considered an undesirable side-effect in size exclusion chromatography. It disrupts the retention sequence determined by molecular size distribution of the components. In practice, a salt (or solvent) solution is used as the mobile phase to reduce ion-exchange (or reversed-phase) binding of solutes. Ironically, the size exclusion becomes as a side-effect in other forms of chromatography, such as affinity or adsorption chromatography.

In this book, the saturation capacities are based on the molar amount of solutes per unit volume of particle skeleton. The differences in saturation capacities can be caused by size exclusion or by sample solutes' chiral discrimination of binding sites. Uneven saturation capacities caused by size exclusion or other reasons bring serious complications in mathematical modeling. The topic of uneven saturation capacities deserves special attention because they are not uncommon. The multicomponent Langmuir isotherm is widely used because of its simplicity and lack of other choices. Unfortunately, it violates the Gibbs-Duhem law of thermodynamics if the saturation capacities are not even [5, 92]. Although it can be considered only as an experimental expression used for correlation, it may not be used for extrapolation over a wider concentration range [5].

In this chapter an isotherm system has been presented for multicomponent systems with uneven saturation capacities induced either by size exclusion or by chiral discrimination of binding sites. The crossover of the isotherms has been studied. A general rate model has been used with the new isotherm system to demonstrate the "peak reversal" phenomenon in multicomponent elution and the crossover of breakthrough curves.

## 7.2   Kinetic and Isotherm Models

A novel mathematical treatment is presented here for adsorption systems with uneven saturation capacities due to size exclusion. It is assumed that one molecule can occupy only one binding site and its binding or size exclusion does not block the availability of other vacant binding sites. That one molecule can only take one binding site is a reasonable assumption for affinity chromatography involving low density immobilized ligands.

Based on these basic assumptions Eq. (3-33) can be modified to give the following kinetic expression that is going to be coupled with the particle phase governing equation:

$$\frac{\partial C_{pi}^*}{\partial t} = k_{ai} C_{pi} \left( C_i^\infty - \sum_{j=1}^{Ns} \theta_{ij} C_{pj}^* \right) - k_{di} C_{pi}^* \tag{7-1}$$

where constants $0 \leq \theta_{ij} \leq 1$ are named "discount factors". They are used to discount the values of $C_{pj}^*$ that belong to the components with a lower degree of size exclusion when doing the calculation of a current component, i.e.,

$$\theta_{ij} = \begin{cases} 1 & i = j \quad \text{or} \quad C_i^\infty \geq C_j^\infty \\ <1 & C_i^\infty < C_j^\infty \end{cases} . \tag{7-2}$$

Equation (7-2) means that, when calculating the $\partial C_{pi}^* / \partial t$ value in Eq. (7-1) for a component with a large molecular size, the $C_{pj}^*$ values for smaller solutes should be discounted, since some of the binding sites for the small solutes are not available to the large solute anyway. Recalling an earlier definition, $F_i^{ex} = \varepsilon_{pi}^a / \varepsilon_p$, in Chap. 3, it is obvious that for component $i$ with a higher degree of size exclusion, its $F_i^{ex}$ value is smaller, and so is its saturation capacity $C_i^\infty$. One may reasonably assume that $\theta_{ij} = C_i^\infty / C_j^\infty$ for those $\theta_{ij}$ values that are apparently not equal to unity. If those $\theta_{ij}$ values are obtained from experimental correlations, the model then becomes semi-empirical.

If an adsorption equilibrium is assumed, Eq. (7-1) becomes

$$b_i C_{pi} \left( C_i^\infty - \sum_{j=1}^{Ns} \theta_{ij} C_{pj}^* \right) - C_{pi}^* = 0 . \tag{7-3}$$

Rearrangement gives

$$b_i C_{pi} C_i^\infty - \sum_{j=1}^{Ns} b_i C_{pi} \theta_{ij} C_{pj}^* - C_{pi}^* = 0 \quad . \tag{7-4}$$

Equation (7-4) can be rewritten in a matrix form shown below,

$$[\mathbf{A}] - [\mathbf{B}][\mathbf{C}_p^*] - [\mathbf{C}_p^*] = 0 \tag{7-5}$$

which gives the following *extended multicomponent Langmuir isotherm:*

$$[\mathbf{C}_p^*] = ([\mathbf{B}] + [\mathbf{I}])^{-1}[\mathbf{A}] \tag{7-6}$$

where

$$A_i = b_i C_{pi} C_i^*, \quad B_{ij} = b_i C_{pi} \theta_{ij}, \quad \text{and} \quad I_{ij} = \begin{cases} 1 & i = j \\ 0 & \text{else} \end{cases} .$$

For a binary system in which component 1 has a higher degree of size exclusion than component 2, one obtains $\theta_{11}=\theta_{22}=\theta_{21}=1$ and $\theta_{12}<1$. The extended binary Langmuir isotherm becomes,

$$C_{p1}^* = \frac{b_1 C_{p1}[(1+b_2 C_{p2})C_1^\infty - \theta_{12} b_2 C_{p2} C_2^\infty]}{1 + b_1 C_{p1} + b_2 C_{p2} + (1-\theta_{12})b_1 C_{p1} b_2 C_{p2}} \tag{7-7}$$

$$C_{p2}^* = \frac{b_2 C_{p2}[(1+b_1 C_{p1})C_2^\infty - b_1 C_{p1} C_1^\infty]}{1 + b_1 C_{p1} + b_2 C_{p2} + (1-\theta_{12})b_1 C_{p1} b_2 C_{p2}} \quad . \tag{7-8}$$

It is obvious that the above two isotherm expressions reduce to the common Langmuir isotherm expressions if $C_1^\infty = C_2^\infty$ and $\theta_{12}=1$. The extended binary Langmuir isotherm has only one extra constant $\theta_{12}$ apart from $C_1^\infty \neq C_2^\infty$ compared with the common Langmuir isotherm. $\theta_{12}$ may be reasonably set to $C_1^\infty / C_2^\infty$. For systems with more than two components, the determination of $\theta_{ij}$ values may not be that simple.

In some multicomponent systems, uneven saturation capacities do not arise from different degrees of size exclusion, but they are induced by an adsorption mechanism at the molecular level. For example, for a binary system, suppose the binding sites (or ligands) are a chiral mixture and they make no difference to component 2, but only some of them are active and usable for component 1; thus component 1 has a lower saturation capacity than component 2. The mathematical treatment for such a system with a chiral discrimination of binding sites is the same as that for systems with uneven saturation capacities that are induced by size exclusion.

## 7.3  Isotherm Crossover

With uneven saturation capacities, an isotherm "crossover" may occur. In this chapter, the isotherm concentration crossover point $C_p^c$ is defined as the concentration in the stagnant fluid inside macropores for a pair of components, at which the corresponding solid phase concentrations ($C_{pi}^*$) are equal.

The concentration crossover point for the binary isotherms given by Eqs. (7-7) and (7-8), can be derived by subtracting the two isotherm expressions and setting $C_{p1} = C_{p2} = C_p^c$, which gives,

$$C_{p1}^* - C_{p2}^* = \frac{(b_1 C_1^\infty - b_2 C_2^\infty)C_p^c + b_1 b_2[2C_1^\infty - (1+\theta_{12})C_2^\infty](C_p^c)^2}{1 + b_1 C_p^c + b_2 C_p^c + (1-\theta_{12})b_1 C_p^c b_2 C_p^c} \quad . \tag{7-9}$$

Setting the left hand side of Eq. (7-9) to zero, one obtains

$$0 = (b_1 C_1^\infty - b_2 C_2^\infty)C_p^c + b_1 b_2 (C_p^c)^2 [2C_1^\infty - (1+\theta_{12})C_2^\infty] \tag{7-10}$$

which gives a nontrivial solution

$$C_p^c = \frac{b_1 C_1^\infty - b_2 C_2^\infty}{b_1 b_2[(1+\theta_{12})C_2^\infty - 2C_1^\infty]} \quad . \tag{7-11}$$

If $\theta_{12} = C_1^\infty / C_2^\infty$,

$$C_p^c = \frac{b_1 C_1^\infty - b_2 C_2^\infty}{b_1 b_2 (C_2^\infty - C_1^\infty)} \quad . \tag{7-12}$$

The denominator of Eq. (7-12) is positive since $C_1^\infty > C_2^\infty$. Thus the binary isotherm has a crossover point if and only if the crossover concentration has a positive value, which requires,

$$b_1 C_1^\infty > b_2 C_2^\infty \quad \text{or} \quad b_1/b_2 > C_2^\infty / C_1^\infty \quad . \tag{7-13}$$

An isotherm concentration crossover signals a selectivity change. The selectivity crossover point may be defined as the critical concentration $C_p^c = C_{p1} = C_{p2}$, which is obtained by setting the relative selectivity of the two components,

$$\text{relative selectivity} = \frac{\partial C_{p1}^* / \partial C_{p1}}{\partial C_{p2}^* / \partial C_{p2}} \tag{7-14}$$

to unity. This gives Eq. (7-15),

$$\partial C_{p1}^* / \partial C_{p1} - \partial C_{p2}^* / \partial C_{p2} = 0 \tag{7-15}$$

from which the following critical selectivity crossover concentration can be easily obtained from Eqs. (7-7) and (7-8) with $C_{p1} = C_{p2} = C_p^c$ and $\theta_{12} = C_1^\infty / C_2^\infty$:

$$C_p^c = -\frac{1}{b_1} + \frac{1}{b_1} \sqrt{\frac{C_1^\infty (b_1 - b_2)}{b_2 (C_2^\infty - C_1^\infty)}} \quad . \tag{7-16}$$

In Eq. (7-16), $C_p^c > 0$ only if $C_1^\infty (b_1 - b_2) / [b_2 (C_2^\infty - C_1^\infty)] > 1$. This leads to Eq. (7-13). Thus, both the concentration crossover and the selectivity crossover require the satisfaction of Eq. (7-13).

It has been known that selectivity depends on the concentration range, and selectivity reversal may occur in the operational concentration range [47]. A selectivity reversal may cause the reversal of the sequence of elution peaks since the migration speed of a component is primarily determined by the $\partial C_{pi}^* / \partial C_{pi}$ value [7].

Figure 7.1 shows two simulated binary elution cases in which component 1 has a smaller saturation capacity and a higher adsorption equilibrium constant than component 2. Parameter values used for the simulation are listed in Table 7.1. The simulation was carried out using the Fortran 77 code KINETIC.F with a simple modification to account for $\theta_{ij}$ in Eq. (7-1), which is not in Eq. (3-33). For the binary system used to discuss the isotherm crossover above, $\theta_{12}$ is set to $C_1^\infty / C_2^\infty$, and $\theta_{11} = \theta_{22} = \theta_{21} = 1$. In the input data file for the computer code, a large value (such as 1000) is assigned to $\mathrm{Da}_i^a$, and a value is then given to $\mathrm{Da}_i^d$ to yield the appropriate values for $b_i$ and $a_i$ according to relationships: $b_i C_{0i} = \mathrm{Da}_i^a / \mathrm{Da}_i^d$ and $a_i = C_i^\infty b_i$. This code will then produce the same results as one using the extended Langmuir isotherm expressed by Eqs. (7-7) and (7-8). The Fortran 77 code RATE.F is not modified to account for $\theta_{ij}$ and used here, since it does not consider the size exclusion effect needed for the discussion below.

The dashed lines in Fig. 7.1 show that component 2 has a smaller retention time than component 1 when the feed concentration of component 1 is low. The solid lines show that component 2 has a higher retention time when the feed concentration of component 1 is increased ten-fold. In Fig. 7.1 (solid lines) the tail end of the component 1 peak is behind that of the component 2 peak. Apparently, at low concentrations component 1 is retained longer than component 2.

Figure 7.2 has the same conditions as Fig. 7.1, except that in the solid line case the concentrations of components 1 and 2 are both 2.0 mol/l in Fig. 7.2. The peak reversal phenomenon is also present in Fig. 7.2. If the uneven saturation capacities are not induced by chiral discrimination of binding sites, but induced by size exclusion, peak reversal can still be present. Figure 7.3 clearly shows such a case in which component 1 has a size exclusion factor of $F_i^{ex}$.

In Figs. 7.1-3, the sample size is quite large ($\tau_{imp} = 1.0$) such that the sample is not diluted too much during its migration inside the column. Otherwise, the

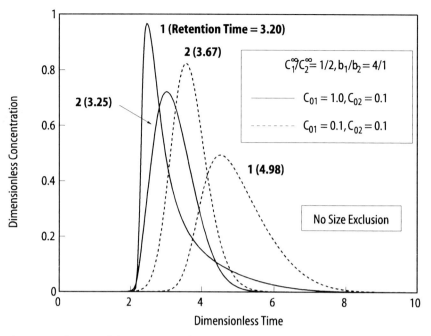

**Fig. 7.1.** Peak reversal due to increased component *1* concentration

dilution of the sample will quickly move the general working concentration range in the isotherm from that over the isotherm crossover point to that below the point. In such a case, peak reversal may not occur at all. Figure 7.4 has the same conditions as Fig. 7.3, except that in Fig. 7.4, the sample size is much smaller ($\tau_{imp}=0.05$). There is no peak reversal in Fig. 7.4 (solid lines) because the concentrations of the two sample components are below the

**Table 7.1.** Parameter values used for simulation in Chap. 7[*]

| Figure(s) | Species | Physical Parameters | | | | | Numerical Parameters | |
|---|---|---|---|---|---|---|---|---|
| | | $Pe_{Li}$ | $\eta_i$ | $Bi_i$ | $a_i$ | $b_i \times C_{0i}$ | Ne | N |
| 7.1-7.4 | 1 | 300 | 4 | 20 | 4 | $4\times$ | 12-22 | 2 |
| | 2 | 300 | 4 | 20 | 2 | $1\times$ | | |
| 7.5 | 1 | 300 | 1 | 40 | 4 | $4\times$ | 8 | 2 |
| | 2 | 300 | 1 | 40 | 2 | $1\times$ | | |
| 7.6 | 1 | 300 | 1 | 20 | 4 | $4\times$ | 8 | 2 |
| | 2 | 300 | 1 | 20 | 1 | $0.5\times$ | | |
| 7.7 | 1 | 300 | 1 | 20 | 4 | $4\times$ | 10 | 2 |
| | 2 | 300 | 4 | 20 | 1 | $0.5\times$ | | |

[a] In all cases, $\varepsilon_b=\varepsilon_p=0.4$. For all elution cases, sample sizes are: $\tau_{imp}=1.0$, except for Fig. 7.4, in which $\tau_{imp}=0.05$.

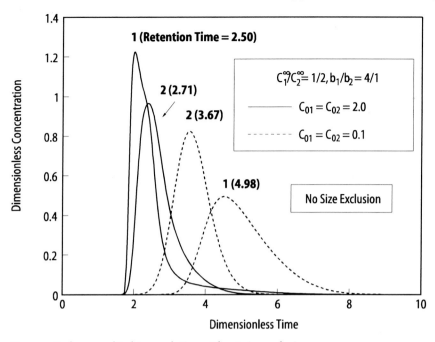

**Fig. 7.2.** Peak reversal in binary elution without size exclusion

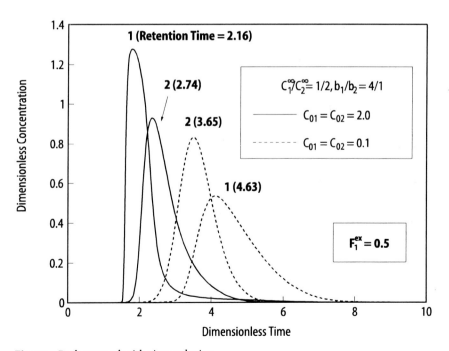

**Fig. 7.3.** Peak reversal with size exclusion

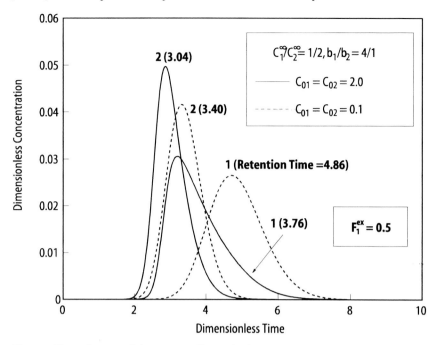

**Fig. 7.4.** No peak reversal due to a small sample size

isotherm crossover point most of the time during their migration inside the column. Their concentrations are quickly diluted after the initial stage of migration because of the small sample size.

The selectivity reversal is also interesting in frontal adsorptions. The solid lines in Fig. 7.5 show that the breakthrough curves cross over each other when the feed concentrations are high. Figure 7.6 (with size exclusion) shows a crossover of breakthrough curves (solid lines). The crossover of breakthrough curves depends not only on the isotherm characteristics and feed concentration, but also on mass transfer conditions. Figure 7.7 has the same conditions as Fig. 7.6, except that $\eta_2=4$ (for component 2) in Fig. 7.7 instead of $\eta_2=1$ in Fig. 7.6. There is a reversal of sequence of breakthrough curves when the feed concentrations are increased, but there is no crossover of the two curves. The absence of a crossover of the two breakthrough curves is apparently because of the change of relative positions of the concentration fronts of the two components arising from a change in mass transfer conditions for component 2.

A detailed treatment of peak reversals due to the isotherm selectivity crossover is considerably more difficult, and involves complicated arguments. A peak reversal is not necessarily the consequence of a selectivity reversal, although a selectivity reversal facilitates a peak reversal.

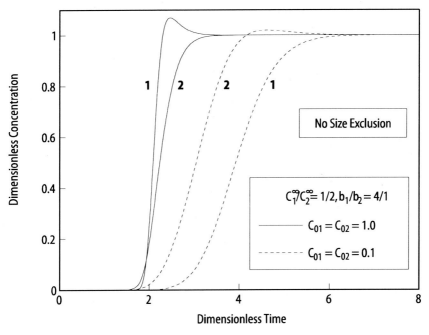

**Fig. 7.5.**   Crossover of breakthrough curves (no size exclusion)

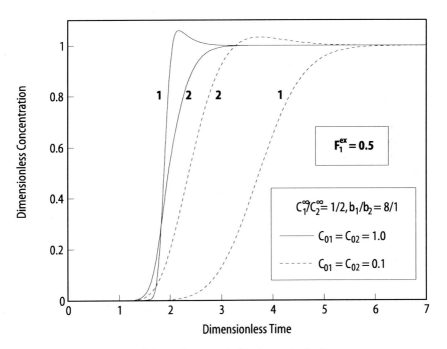

**Fig. 7.6.**   Crossover of breakthrough curves (with size exclusion)

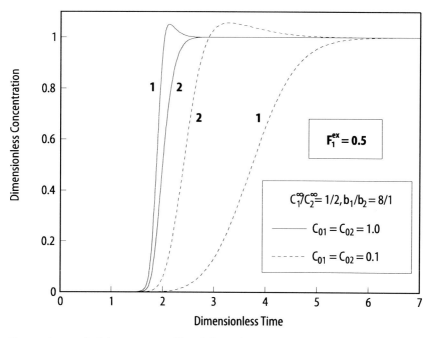

**Fig. 7.7.** Reversal of the sequence of breakthrough curves without crossover of the curves

## 7.4  Summary

An isotherm system has been presented for multicomponent systems with uneven saturation capacities induced either by size exclusion or chiral discrimination of the binding sites. The mathematical criteria for isotherm crossover in terms of concentration and selectivity have been derived and discussed for the isotherm. The isotherm system serves as a valuable model for experimental correlation of isotherm data showing uneven saturation capacities. Using a general rate model that counts for an extended binary Langmuir isotherm system, the peak reversal phenomenon in elutions, and the crossover of breakthrough curves in frontal adsorptions has been demonstrated. The methodology demonstrated here for systems with uneven saturation capacities can be readily applied to common stoichiometric ion-exchange systems.

# 8 Modeling of Affinity Chromatography

## 8.1 Introduction

Affinity chromatography has seen rapid growth in recent years. It is a powerful tool for the purification of enzymes, antibodies, antigens, and many other proteins and macromolecules that are of important use in scientific research and development of novel biological drugs. Affinity chromatography not only purifies a product, but also concentrates the product to a considerable extent [32]. Over the years, this subject has been reviewed by many people, including Chase [32], and Liapis [93]. Affinity chromatography is also called biospecific adsorption, since it utilizes the biospecific binding between the solute molecules and immobilized ligands. The monovalent binding between a ligand and a solute macromolecule is generally considered as following second order kinetics expressed by Eqs. (3-32) and (3-33).

A class of monoclonal antibody used in affinity chromatography is immunoglobin G, which has two identical antigen binding sites. If the binding of one antigen does not interfere with the binding of another antigen onto the other binding site of the same antibody, then bindings can be considered as two monovalent bindings. If the antigen has more than one binding site that can be recognized by the antibody, multivalent bindings are possible. This was discussed by Chase [32].

There are two kinds of bindings in affinity chromatography, specific and non-specific. The specific binding involves only the target macromolecule and the ligand. Non-specific binding is an undesirable, but often unavoidable side-effect. It can be caused by unintended ion-exchange or hydrophobic interaction.

The operational stages of affinity chromatography include adsorption, washing and elution. The column is regenerated after each cycle. The adsorption stage is carried out in the form of a frontal adsorption. In order to obtain a sharp concentration front for the target macromolecule, a small flow rate is often used [32]. The washing stage right after the adsorption stage is aimed at removing the impurities in the bulk fluid and in the stagnant fluid inside particle macropores, and impurities bonded to the stationary phase via non-specific binding [32].

The elution stage removes the bonded target macromolecules from the ligands. Elution can be carried out by using a soluble ligand that is often the same as that immobilized in the stationary phase, provided that the soluble ligand is present in a higher concentration and is relatively inexpensive. The other method is called non-specific desorption, which uses a variety of eluting agents, such as pH, protein denaturants, chaotropic agents, polarity reducing agents, temperature [32] to weaken the binding between the macromolecules and immobilized ligands.

Elution in affinity chromatography has a different meaning from that used in other forms of chromatography, such as reversed phase and ion-exchange, in which elution means impulse analysis. To avoid confusion, impulse analysis in affinity chromatography is referred as zonal analysis [31, 36, 63].

The Langmuir isotherm for biospecific binding, which comes from second order kinetics, is characterized by a very large equilibrium constant ($b$), and a very small saturation capacity ($C^\infty$), indicating that the ligand density of an affinity matrix is often quite low. Because of the large equilibrium constant, the isotherm can be nonlinear even if the concentration of macromolecules is very low. A universal function was introduced by Lee et al. [63] to measure the effect of isotherm nonlinearity in zonal analysis for affinity chromatography.

General rate models were developed by Arve and Liapis [94, 95] for affinity chromatography. Their models consider various mass transfer mechanisms and second order kinetics between the immobilized ligands and the macromolecules, and between the soluble ligands and the macromolecules during elution.

## 8.2   Effect of Reaction Kinetics

A general rate model with second order kinetics has already been described in Chap. 5. The corresponding Fortran 77 code included in the diskette accompanying this book is named KINETIC.F. The code can be used to study kinetic effects. The three breakthrough curves in Fig. 8.1 show the effect of reaction rates. Parameter values used for the simulation are listed in Table 8.1. The solid curve shows, that when the Damkölher numbers for binding and dissociation are low, the breakthrough curve takes off sharper and earlier. It slowly levels off later. This is due to slow reaction rates.

Fig. 8.2 shows that when the reaction rates increase to some extent the breakthrough curves will be very close to that of the equilibrium case. In fact, the equilibrium case is the asymptotical limit of the results obtained from the kinetic model with large Damkölher numbers.

Fig. 8.3 shows the effect of reaction rates in a single component zonal elution case. The solid line shows that the elution peak appears early with a very sharp front, but it has a very long tail. This indicates that when the reaction rates are very low, a large portion of the solute molecules do not have a chance to bind with the ligands and they are eluted out quite quickly. On the other hand, those molecules that do bind with the ligands are dissociated very

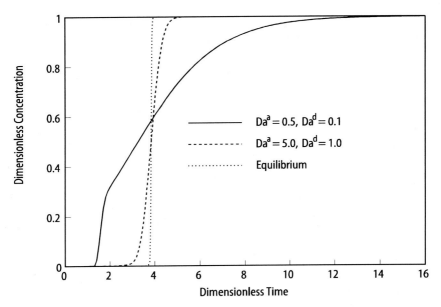

**Fig. 8.1.** Effect of reaction rates in frontal analysis

**Table 8.1.** Parameter values used for simulation in Chap. 8[*]

| Figure(s) | Species | Physical Parameters | | | | | | | Numerical Parameters | |
|---|---|---|---|---|---|---|---|---|---|---|
| | | $Pe_{Li}$ | $\eta_i$ | $Bi_i$ | $C^\infty$ | $C_{0i}/C_{01}$ | $Da_i^a$ | $Da_i^d$ | Ne | N |
| 8.1 | 1 | 400 | 20 | 10 | 0.3 | | 0.5 | 0.1 | 8 | 1 |
| 8.2 | 1 | 400 | 20 | 10 | 0.3 | | 5 | 1 | 8 | 1 |
| 8.3 | 1 | 400 | 10 | 20 | 1 | | 1 | 0.2 | 13 | 2 |
| 8.4 | 1 | 400 | | | 0.3 | | 0.5 | 0.1 | 8 | 1 |
| 8.5 | 1 | 400 | 0.1 | 10 | 0.3 | | 50 | 10 | 9 | 2 |
| 8.6 | 1 | 400 | 0.1 | 5 | | | | | 10 | 2 |
| 8.6 (dash) | 1 | 400 | 10 | 10 | 1 | | 0.5 | 0.1 | 8 | 2 |
| 8.7 | 1 | 400 | | | | | | | 10 | 2 |
| 8.8 | 1 | 400 | 20 | 10 | 0.3 | | 0.5 | 0.1 | 8 | 1 |
| 8.9 | 1 | 200 | 2 | 10 | | | | | 10 | 1 |
| 8.10 | 1 | 300 | 10 | 40 | 10 | | 2 | 0.2 | 7 | 2 |
| 8.11 | 1 | 300 | 10 | 40 | 10 | 1 | 2 | 0.2 | 7 | 2 |
| | 2 | 300 | 10 | 40 | | 1 | 2 | 0.2 | | |
| | 3 | 300 | 10 | 40 | | 1 | | | | |
| 8.12 | 1 | 300 | 10 | 40 | 10 | 1 | 2 | 0.2 | 8 | 2 |
| | 2 | 300 | 10 | 40 | | 5 | 10 | 0.2 | | |
| | 3 | 300 | 10 | 40 | | 1 | | | | |

[*] Parameters are for solid lines in figures except Fig. 8.6. In all cases, $\varepsilon_b=\varepsilon_p=0.4$. In Fig. 8.3, $\tau_{imp}=0.5$. For the solid line case in Fig. 8.6, the parameters for Langmuir isotherm is $a=5$, $bC_0=5$. For Fig. 8.7, $a=5$ in the Langmuir isotherm. In Figs. 8.10 to 8.12, the size exclusion factor for all components is $F_i^{ex}=0.8$. CPU times on SUN 4/280 computer are: Fig. 8.2 (solid line), 9.9 sec; Fig. 8.3 (solid line), 3.8 min; Fig. 8.12, 24.7 min.

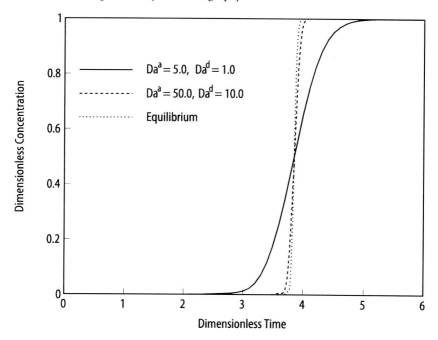

**Fig. 8.2.** Fast reaction rates vs. equilibrium

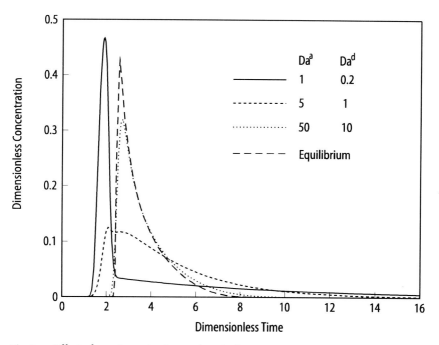

**Fig. 8.3.** Effect of reaction rates in zonal analysis

slowly, causing a long tail. This is partially reflected by the breakthrough curve shown as the solid line in Fig. 8.1 since the two operational modes are interrelated. Figure 8.3 also shows that the peak front appears later, and the peak height reduces when the reaction rates increase. When the reaction rates further increase, the appearance of the peak front is delayed even more, and the peak height increases. The increase of reaction rates reduces the tailing effect and sharpens the peak front (see Fig. 8.3).

In Fig. 8.4 the solid curve is the same as that in Fig. 8.1. Figure 8.4 shows that the slow reaction rates are the rate-limiting step in this case. On the other hand, Fig. 8.5 shows a case in which the mass transfer rates are rate-limiting, since the reaction rates in the figure are relatively much higher than the mass transfer rates.

The breakthrough curve of an adsorption system with slow mass transfer rates is similar to that of slow reaction rates. Both take off sharper and earlier and then slowly level off later. In Fig. 8.6, one may find that the two cases differ in a revealing way. For the slow mass transfer case, the breakthrough curve takes off earlier (at $\tau<1$) than in the case with slow reaction rates, since, in the former case, many solutes do not enter the particles, while in the latter case they do. The slow mass transfer here means that both the external film mass transfer and intraparticle diffusion rates are low. The solid line in Fig.

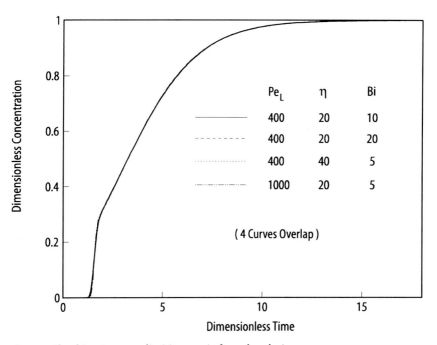

| | $Pe_L$ | $\eta$ | $Bi$ |
|---|---|---|---|
| ———— | 400 | 20 | 10 |
| - - - - - - | 400 | 20 | 20 |
| ············· | 400 | 40 | 5 |
| —·—··—··— | 1000 | 20 | 5 |

( 4 Curves Overlap )

**Fig. 8.4.** Slow kinetics as rate limiting step in frontal analysis

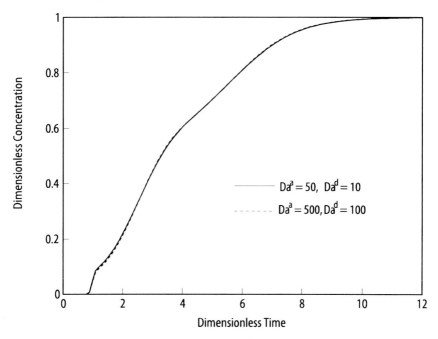

**Fig. 8.5.** Mass transfer as rate limiting step in frontal analysis

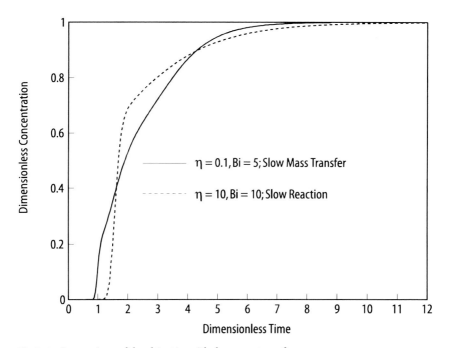

**Fig. 8.6.** Comparison of slow kinetics with slow mass transfer

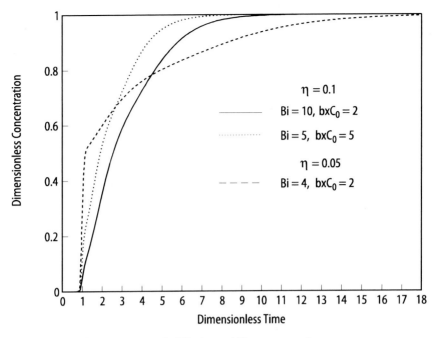

**Fig. 8.7.**  Effect of slow intraparticle diffusion and film mass transfer

8.7 shows that the take-off of the breakthrough curve is very sharp if the film mass transfer coefficient is small even though the intraparticle diffusion coefficient is not, since many solutes do not have a chance to penetrate the liquid film into the macropores of the particles.

## 8.3  Effect of Size Exclusion

The effect of size exclusion on the adsorption saturation capacity has been discussed in Chap. 7. The reduction of the column hold-up capacity due to the effect of size exclusion is clearly shown in Fig. 8.8. The two single-component breakthrough curves have the same conditions except that the dashed line case has a size exclusion effect and half of the particle porosity is inaccessible, i.e., $F^{ex} = \varepsilon_p^a / \varepsilon_p = 0.5$. The capacity in the size exclusion case has been set to half of that without size exclusion (solid line). Figure 8.8 shows that in the case of size exclusion, the breakthrough curve tends to be sharper. This is also true for systems with no adsorption as shown in Fig. 8.9. In both Figs. 8.8 and 8.9, the column hold-up capacity area can be checked using Eq. (3-39). The dotted curve in Fig. 8.9 shows a case with a complete size exclusion. The area above

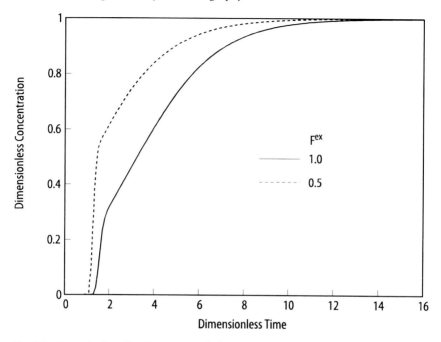

**Fig. 8.8.** Size exclusion effect in presence of adsorption

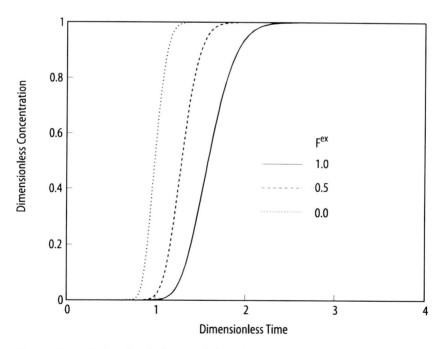

**Fig. 8.9.** Size exclusion effect in absence of adsorption

the curve is found to be unity. One may expect this kind of case when using a totally nonpenetrating and nonadsorbing substance such as blue dextrin to measure the bed void volume fraction.

## 8.4 Interaction Between Soluble Ligand and Macromolecule

Soluble ligands can be used to elute the adsorbed macromolecules in the elution stage if the ligands are not expensive and can be easily separated from the macromolecules after elution [32]. A rate model involving soluble ligand used for the elution of a single adsorbate was reported by Arve and Liapis [39, 94] for finite bath and fixed-bed operations.

### 8.4.1 Modeling of Reaction in the Fluid

The kinetic rate model described in Chap. 3 can be extended to include a binding reaction in the bulk fluid and the stagnant fluid inside macropores of the particles between macromolecule $P$ (component 1) and soluble ligand $I$ (component 2). The complex formed from the binding of $P$ and $I$ is $PI$ (component 3).

$$P+I \underset{k_{d2}}{\overset{k_{a2}}{\rightleftharpoons}} PI \ . \tag{8-1}$$

In Eq. (8-1), $k_{a2}$ and $k_{d2}$ are the association and dissociation constants for $P$ and $I$. The binding between the macromolecule and the immobilized ligand L forms PL,

$$P+L \underset{k_{d1}}{\overset{k_{a1}}{\rightleftharpoons}} PL \ . \tag{8-2}$$

It is assumed that each macromolecule can bind with only one ligand, $I$ or $L$, and there is no interaction between the two different ligands, $I$ and $L$.

Size exclusion effect in a system that involves large molecules such as $P$, $PL$ and possibly $L$, may not be negligible. If so, they should be included in the model. The kinetic rate model described in Chap. 3 can be readily extended to accommodate this. For simplicity, only a three-component system is discussed here, since a generalized system in this case is awkward to present.

The extension of the kinetic rate model can be carried out as follows.

(1) Bulk-Fluid Phase Governing Equation

$$-D_{bi}\frac{\partial^2 C_{bi}}{\partial Z^2}+v\frac{\partial C_{bi}}{\partial Z}+\frac{\partial C_{bi}}{\partial t}+\frac{3k_i(1-\varepsilon_b)}{\varepsilon_b R_p}(C_{bi}-C_{pi,R=R_p})$$
$$-f(i)(k_{a2}C_{b1}C_{b2}-k_{d2}C_{b3})=0 \tag{8-3}$$

where $f(i)=-1$ is for components 1 and 2 ($i=1,2$), and $f(i)=1$ for component 3 ($i=3$).

(2)  Particle Phase Governing Equation

$$g(i)(1-\varepsilon_p)\frac{\partial C^*_{pi}}{\partial t}+\varepsilon^a_{pi}\frac{\partial C_{pi}}{\partial t}-\varepsilon^a_{pi}D_{pi}\left[\frac{1}{R^2}\frac{\partial}{\partial R}\left(R^2\frac{\partial C_{pi}}{\partial R}\right)\right]$$
$$-f(i)\,\varepsilon^a_{pi}(k_{a2}C_{p1}C_{p2}-k_{d2}C_{p3})=0 \tag{8-4}$$

$$\partial C^*_{p1}/\partial t=k_{a1}C_{p1}(C^\infty_1-C^*_{p1})-k_{d1}C^*_{p1} \tag{8-5}$$

in which $g(i)=1$ for $i=1$, and $(i)=0$ for $i=2,3$, since only component 1 binds with the immobilized ligand. The use of sign changers, $f(i)$ and $g(i)$, is purely for the compactness of the model system in its written form. Note that $C^*_{p1}$ represents concentration $[PL]$.

The model system presented here is more general than a similar one presented by Arve and Liapis [94] since size exclusion is included. This model seems to be more like a model for a fixed-bed reactor than for chromatography, because there is a new component $(PI)$ forming and leaving the column. Defining the following dimensionless constants:

$$c_{bi}=C_{bi}/C_{0i},\quad c_{pi}=C_{pi}/C_{0i},\quad c^*_{pi}=C^*_{pi}/C_{0i},\quad c^\infty_1=C^\infty_1/C_{01},\quad r=R/R_p$$

$$z=Z/L,\quad \mathrm{Pe}_{Li}=vL/D_{bi},\quad \mathrm{Bi}_i=k_iR_p/(\varepsilon^a_{pi}D_{pi}),\quad \eta_i=\varepsilon^a_{pi}D_{pi}L/(R_p^2 v)$$

$$\xi_i=3\mathrm{Bi}_i\eta_i(1-\varepsilon_b)/\varepsilon_b,\quad \tau=vt/L,\quad \mathrm{Da}^a_1=L(k_{a1}C_{01})/v,\quad \mathrm{Da}^d_1=Lk_{d1}/v$$

$$\mathrm{Da}^a_2=L(k_{a2}C_{01})/v,\quad \mathrm{Da}^d_2=Lk_{d2}/v$$

the PDE system can be expressed in dimensionless forms as follows:

$$-\frac{1}{\mathrm{Pe}_{Li}}\frac{\partial^2 c_{bi}}{\partial z^2}+\frac{\partial c_{bi}}{\partial z}+\frac{\partial c_{bi}}{\partial \tau}+\xi_i(c_{bi}-c_{pi,r=1})$$
$$-f(i)\left(\mathrm{Da}^a_2 c_{b1}\frac{C_{02}}{C_{0i}}c_{b2}-\mathrm{Da}^d_2\frac{C_{03}}{C_{0i}}c_{b3}\right)=0 \tag{8-6}$$

$$g(i)\frac{\partial}{\partial\tau}(1-\varepsilon_p)c^*_{pi}+\varepsilon^a_{pi}\frac{\partial c_{pi}}{\partial\tau}-f(i)\varepsilon^a_{pi}\left(\mathrm{Da}^a_2 c_{p1}\frac{C_{02}}{C_{0i}}c_{p2}-\mathrm{Da}^d_2\frac{C_{03}}{C_{0i}}c_{p3}\right)$$
$$-\eta_i\left[\frac{1}{r^2}\frac{\partial}{\partial r}\left(r^2\frac{\partial c_{pi}}{\partial r}\right)\right]=0 \tag{8-7}$$

$$\frac{\partial c^*_{p1}}{\partial\tau}=\mathrm{Da}^a_1 c_{p1}(c^\infty_1-c^*_{p1})-\mathrm{Da}^d_1 c^*_{p1}\quad. \tag{8-8}$$

Since $C_{03}$ is not known before simulation, it is replaced by $C_{01}$ for the nondimensionalization of the concentrations of component 3 such that $C_{b3}=c_{b3}C_{01}$ and $C_{p3}=c_{p3}C_{01}$.

### 8.4.2 Solution Strategy

The numerical procedure described in Chap. 3 for the kinetic model can be modified to implement the fluid phase reaction. For the bulk-fluid phase, the finite element vector $(\mathbf{AFB_i})_m^e$ should now include the last term of Eq. (8-6) and Eq. (8-9) should be used to replace Eq. (3-27):

$$(\mathbf{AFB_i})_m^e = \int \left[ \xi_i \phi_m c_{pi,r=1} \right.$$
$$\left. + \phi_m f(i) \left( Da_2^a c_{b1} \frac{C_{02}}{C_{0i}} c_{b2} - Da_2^d \frac{C_{03}}{C_{0i}} c_{b3} \right) \right|_e \right] dz \quad . \tag{8-5}$$

The modification of the particle phase governing equation is quite straightforward. Details are omitted here. The Fortran 77 code for the numerical solution to the new model named "AFFINITY.F" is available.

## 8.5 Modeling of the Three Stages in Affinity Chromatography

Fig. 8.10 shows an affinity chromatographic separation with a wash stage after the frontal adsorption stage. The non-specifically bound impurities are not included in the simulation. Their effluent histories can be simulated in a separate run and then superimposed onto the current figure, since they do not interact with the macromolecule. Because no soluble ligand or other active eluting agent is used for the elution, the simulated chromatogram shows a very long tail. It indicates that the recovery of the macromolecule is difficult and not efficient.

Fig. 8.11 has the same condition as Fig. 8.10, except that soluble ligands are used for elution at $\tau=15$ after the wash stage that started at $\tau=14$ and ended at $\tau=15$. Compared with Fig. 8.10, it is obvious that elution using soluble ligands helps reduce tailing and the time needed for the recovery of the product. If a higher concentration of soluble ligand is used for elution, the elution stage will be shorter, and the recovered product will have a higher concentration. This is demonstrated by Fig. 8.12, in which the soluble ligand concentration in the feed is five times that in Fig. 8.11.

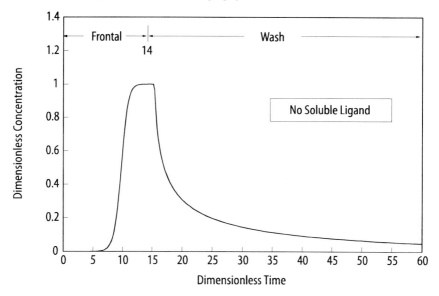

**Fig.8.10.** Frontal adsorption stage combined with wash stage in affinity chromatography

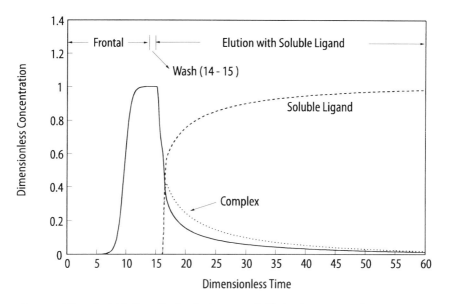

**Fig.8.11.** Effect of soluble ligand in the elution stage of affinity chromatography

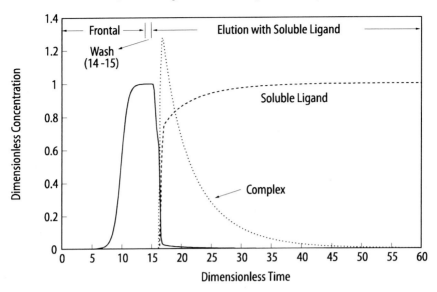

**Fig.8.12.** Effect of soluble ligand concentration in elution

## 8.6 How to Use the Fortran 77 Code AFFINITY.F

Below is the heading generated by the code AFFINITY.F upon its execution with the data file for Fig. 8.12.

```
Affinity Chromatography (w/ Soluble Ligand) Simulator by Tingyue Gu (Ohio
U.)
======================================================================
nsp nelemb nc index  timp   tint  tmax  epsip  epsib
 3   12     2    4   14.000  0.200  60.0  0.400  0.400

    PeL     eta      Bi    Cinf     CO     Daa     Dad    exf
  300.00  10.000  40.000 0.10E-04 0.10E-05  2.000   0.200 0.800E+00
  300.00  10.000  40.000 0.00E+00 0.50E-05 10.000   0.200 0.800E+00
  300.00  10.000  40.000 0.00E+00 0.10E-05  0.000   0.000 0.800E+00
 Switch to displacement at tshift =  15.000
============================================== End of data file
Total ODE =  375    Data pts =  300

Soluble ligand I. 1+2 <==> 3: P(rotein) + I <==> PI
index =0  Displacement with soluble ligand
index =1  Frontal with two feed streams: P, I
index =2  Strict isocratic elution with I in sample
index =3  Pseudo-isocratic elution, no I in sample
index =4  Frontal (P only) then at tshift displ. w/ I
index =5  Same as =4, but reverse flow at displ.
```

```
Results (t, c1, c2, c3) follow.    Please wait...
0.2000    0.00000    0.00000    0.00000
(...more data points)
```

If the text is all stripped from the heading, the remaining numerical figures from the beginning to "End of data file" is all that contained in the data file. The meaning of input items are the same as those for the Fortran 77 code KINETIC.F. The $\tau_{imp}=14$ ("timp"=14) indicates the duration that the protein (component 1) solution is fed to the column. At $\tau=15$ ("tshift"=15), the column feed is switched to the soluble ligand (component 2) solution for elution. The difference between the two dimensionless times is $\Delta\tau=1$, which is the time for the wash stage. In the output data, component 3 is complex $PI$.

By setting index = 2 or 3, KINETIC.F can be used to simulate a zonal elution with a soluble ligand serving as an inhibitor in the mobile phase. If the injected sample contains the soluble ligand, index = 2. If not, index = 3.

## 8.7  Summary

In this Chapter, various aspects of affinity chromatography, including the effects of reaction kinetics, mass transfer, size exclusion have been discussed using simulation. The kinetic rate model has been modified to describe the reaction between macromolecules and soluble ligands in the bulk-fluid and in stagnant fluid inside particle macropores. The role of soluble ligand in the elution stage in affinity chromatography has been investigated.

# 9 Modeling of Multicomponent Gradient Elution

## 9.1 Introduction

Gradient elution chromatography is a very important method in chromatographic separations, especially for proteins, because they have a wide range of retentivity. In gradient elution, a modulator is used in the mobile phase to adjust eluent strength. The modulator can be acetonitrile in reversed phase chromatography, or ammonium sulfate in hydrophobic interaction chromatography, or sodium chloride in ion-exchange chromatography. In ion-exchange chromatography, a pH gradient may also be used. The modulator concentration in the mobile phase is increased (or decreased as in hydrophobic interaction chromatography) continuously with time. This change in the strength of modulator allows gradient elution to separate components with widely different retentivities. In preparative- or large-scale operations, gradient elution can concentrate a sample while achieving a purification at the same time. In isocratic elution, a sample is always diluted to a certain degree. Because of this, gradient elution is often desired when handling large volumes of sample.

Gradient elution in analytical HPLC involves small and dilute samples separated on a highly efficient column. In preparative- and large-scale gradient elution chromatography the column is often overloaded in terms of sample feed volume and/or concentration, or both. The column may not be a highly efficient column due to scale and cost factors. Larger particle sizes may be used for column packing. Thus, interference effects, axial dispersion and mass transfer resistance such as interfacial film mass transfer and intraparticle diffusion may become important.

The scale-up of protein purification using gradient elution was often carried out empirically [96]. The theoretical basis of gradient elution in nonlinear chromatography has not been well established [47]. Because of the complications involved in the modeling of gradient elution, very few existing models considered mass transfer resistance or kinetic resistance, although some considered axial dispersion [47, 96-98]. Existing mathematical models in the literature for gradient elution were reviewed by Gu et al. [99].

Melander et al. [4] proposed an eluite-modulator relationship, which is suitable for both electrostatic and hydrophobic interactions. This relationship was supported by some thermodynamic arguments. In this chapter, a general rate model for multicomponent elution chromatography is presented. In the model it is assumed that the eluites follow the multicomponent Langmuir isotherm with a uniform saturation capacity, $C^\infty$. The $b_i$ values are a function of the modulator concentration following the eluite-modulator relationship proposed by Melander et al. The model is capable of simulating various gradient operations with linear, nonlinear and stepwise linear gradients.

## 9.2  General Rate Model for Multicomponent Gradient Elution

The modulator is designated as the last component in a multicomponent system, which is component Ns. Thus, the eluite-modulator relationship proposed by Melander et al. [4] can be written as follows:

$$\log_{10} b_i = \alpha_i - \beta_i \log_{10} C_{p,Ns} + \gamma_i C_{p,Ns} \qquad (9\text{-}1)$$

in which $\alpha_i$, $\beta_i$ and $\gamma_i$ are experimental correlation parameters. Note that Melander et al. used retention factor $k_i'$ (also known as the capacity factor) instead of the adsorption equilibrium constant in the Langmuir isotherm, $b_i$. But $k_i' = \phi C^\infty b_i$, and the constant value of $\phi C^\infty$ can be separated from $\phi C^\infty b$ and lumped into to the $\alpha$ term in Eq. (9-1). It is assumed that eluites do not interfere with each other's correlation parameters. The saturation capacities for all the eluites are the same and they are not affected by the modulator concentration. This correlation implies that when the modulator concentration is zero, the $b_i$ values for eluites are infinity, indicating irreversible bindings. The column should be presaturated with a small nonzero modulator concentration (denoted $C_{m0}$).

The model requires the following initial conditions:

at $\tau=0$, $c_{bi}=c_{pi}=c_{pi}^*=0$ $\qquad\qquad$ for the eluites (i=1,2,...,Ns–1), and

at $\tau=0$, $c_{bi}=c_{pi}=C_{m0}/C_{0i}=c_{m0}$, and $c_{pi}^*=0$ for the modulator (i=Ns).

The dimensionless feed concentration profiles for the boundary conditions at the column inlet are as follows:

For the eluites (i=1,2,...,Ns–1),

$$\frac{C_{fi}(\tau)}{C_{0i}} = \begin{cases} 1 & 0 \le \tau \le \tau_{imp} \\ 0 & \tau > \tau_{imp} \end{cases} . \qquad (9\text{-}2)$$

For the modulator (i=Ns),

$$\frac{C_{fi}(\tau)}{C_{0i}} = \begin{cases} = C_{m0}/C_{0i} & -\infty \le \tau \le \tau_{imp} \\ \ge (\text{or}) \le C_{m0}/C_{0i} & \tau > \tau_{imp} \end{cases} . \qquad (9\text{-}3)$$

The upper boundary value of the rectangular sample pulse of an eluite is taken as their reference concentration value, $C_{0i}$. For the modulator, its reference concentration value can take any convenient value, such as the $C_{m0}$ value that is the modulator concentration in the column before a gradient takeoff. The gradient profile of a modulator concentration is described in Eq. (9-3). It can take any shape after a sample injection (i.e., after $\tau > \tau_{imp}$). If the takeoff of the modulator concentration after a sample injection is of a nonlinear nature, i.e., $C_{f,Ns}(\tau)/C_{0,Ns}$ vs $\tau$ is nonlinear for $\tau > \tau_{imp}$, the process becomes a nonlinear gradient elution.

## 9.3  Numerical Solution

The model is numerically solved with a Fortran 77 code named "GRADIENT.F". The code was obtained by modifying the existing code KINETIC.F that was written for the kinetic rate model described in Chap. 3.

The use of the kinetic rate model instead of an equilibrium rate model gives a special advantage in dealing with gradient elution with variable $b_i$ values for the eluites. In the equilibrium rate model, the multicomponent Langmuir isotherm is directly inserted into the particle phase governing equation to eliminate $c_{pi}^*$ in Eq. (3-10). This makes the left hand side of Eq. (3-10) too complicated for the evaluation of time derivatives of particle phase concentrations in gradient elution, since $a_i$ and $b_i$ in the multicomponent Langmuir isotherm are also time dependent variables because of Eq. (9-1). Fortunately, all the complications are not present if the kinetic model is used.

The asymptotic limit of the kinetic model is the equilibrium rate model. To use the kinetic model as an equilibrium rate model for gradient elution, one only has to set the Damköhler number for desorption (or adsorption) of eluites ($i=1,2,...,Ns-1$) in the kinetic model to a large arbitrary value (say, no less than 1000) and then calculate the Damköhler number for adsorption (or desorption) from the relationship $Da_i^a / Da_i^d = b_i C_{0i}$, where $b_i$ is obtained from Eq. (9-1). By doing so, Eq. (9-1) is combined with the kinetic model without any difficulty. The incorporation of initial conditions in the Fortran code required for gradient elution is straightforward.

## 9.4  How to Use the Fortran 77 Code GRADIENT.F

To demonstrate the capability of the Fortran 77 code, GRADIENT.F, two simulation cases are presented below. In both cases the sample size is $\tau_{imp} = 0.3$, and $\varepsilon_p = \varepsilon_b = 0.4$. It is assumed that the adsorption of the modulator onto the stationary phase is negligible. This is done by setting the righthand side of Eq. (3-34) to zero for component Ns.

Figure 9.1 shows a simulated chromatogram of a gradient elution of three components with a modulator (component 4). Parameters used for the simulation are listed in Table 9.1. The eluite-modulator relationships are shown in the figure insert of Fig. 9.1. Two linear gradients are used to achieve a complete baseline separation within a short period of time.

The Fortran 77 code for the simulation of multicomponent gradient elution requires a data file named "data" to supply simulation parameters. Below is the heading of the results obtained for Fig. 9.1.

```
Gradient Elution Simulator by Tingyue Gu (Ohio U.)
==================================================================
nsp nelemb  nc index  timp   tint  tmax  epsip  epsib
 4    12     2    3   0.300  0.010  7.0   0.400  0.400

    PeL       eta      Bi      CO      Cinf      aa       bb       cc
  350.00     4.000  30.000 0.50E-04 0.10E-03  -3.000   2.600    3.200
  350.00     4.000  30.000 0.10E-04 0.10E-03  -3.000   3.000    3.200
  400.00     6.000  20.000 0.10E-04 0.10E-03  -3.000   3.500    3.200
  400.00    10.000   5.000 0.10E+00 0.00E+00   0.000   0.000    0.000
Dimensionless cnsp0 = 0.01500   stl=  0.3000  st2=  3.1000

    a1        a2       a3       a4       a5       a6
 0.01500   0.01500  0.00000  0.00000  0.00000  0.00000
    a11       a22      a33      a44      a55      a66
-0.22300   0.10000  0.00000  0.00000  0.00000  0.00000
============================================ End of data file.
Continuity of two linear gradients requires a11 = -0.2230
Total ODE = 500 data pts = 700

Gradient elution with gradient profile: at t >= st1,
C(t)/CO=a1 +a2*tt +a3*tt*tt +a4*tt**3 +a5*exp(a6*tt)
where tt = t - timp.
Second gradient change starts from t >= st2 with
C(t)/CO=a11+a22*tt+a33*tt*tt+a44*tt**3+a55*exp(a66*tt)
```

Table 9.1. Parameter values used for simulation in Chap. 9[*]

| Figure(s) | Species | Physical Parameters | | | | | Numerical Parameters | |
|---|---|---|---|---|---|---|---|---|
| | | $Pe_{Li}$ | $\eta_i$ | $Bi_i$ | $C_{0i}$ | $C^\infty$ | Ne | N |
| 9.1 | 1 | 350 | 4 | 30 | $5\times10^{-5}$ | $1\times10^{-4}$ | 11 | 2 |
| | 2 | 350 | 4 | 30 | $1\times10^{-5}$ | $1\times10^{-4}$ | | |
| | 3 | 400 | 6 | 20 | $1\times10^{-5}$ | $1\times10^{-4}$ | | |
| | 4 | 400 | 100 | 5 | 0.1 | 0 | | |
| 9.2 | 1 | 350 | 4 | 30 | $5\times10^{-5}$ | $1\times10^{-4}$ | 10 | 2 |
| | 2 | 350 | 4 | 30 | $1\times10^{-5}$ | $1\times10^{-4}$ | | |
| | 3 | 400 | 6 | 20 | $1\times10^{-5}$ | $1\times10^{-4}$ | | |
| | 4 | 400 | 100 | 5 | 0.1 | 0 | | |

[*] Dimensionless initial modulator concentration for both figures is 0.015. The feed gradient profiles $C_{f,Ns}(\tau)/C_{0,Ns}$ are: Fig. 9.1, $0.015+0.015(\tau-\tau_{imp})$ for $\tau_{imp}\leq\tau<3.1$, and $-0.223+0.1(\tau-\tau_{imp})$ for $\tau\geq3.1$; Fig. 9.2, $0.015+0.007(\tau-\tau_{imp})^2$ for $\tau\geq\tau_{imp}$.

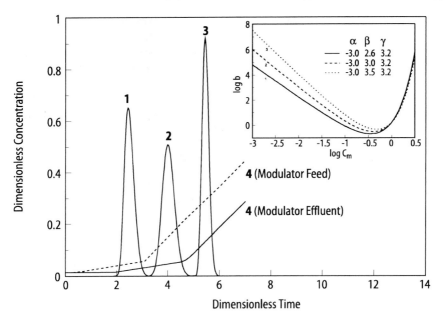

**Fig. 9.1.** Gradient elution with two consecutive linear gradients of the modulator

```
Results (t, c1, c2, c3...) follow. Please wait...

    0.0100 0.00000 0.00000 0.00000 0.01500
    (...more data points)
```

Stripping away the text from the beginning to "End of data file" in the print-out above, the remaining numerical values are what the file "data" contains. The meaning of the symbols are: nsp=Ns, nelemb=Ne, nc=N, index=$\log_{10}$ Da$_i^d$ (a value no less than three in order to convert the kinetic model to the equilibrium rate model), timp=$\tau_{imp}$, tint=$\tau$ interval used in data output, tmax= maximum $\tau$ value used for simulation, epsip=$\varepsilon_p$, epsib=$\varepsilon_b$, PeL=Pe$_{Li}$, eta=$\eta_i$, Bi=Bi$_i$, C0=$C_{0i}$, Cinf=$C^\infty$, aa=$\alpha_i$, bb=$\beta_i$, cc=$\gamma_i$, cnsp0 =dimensionless feed concentration of the modulator, st1=starting time for the first gradient, st2= starting time for the second gradient (if there is no second gradient, st2 should be set to a value larger than tmax), a1 to a6 and a11 to a66 are coefficients in the two gradient profiles. If st2>tmax, the program does not require coefficients a11 to a66 in the input data file.

When $\tau$<st1, the dimensionless modulator feed concentration is automatically maintained at cnsp0. When $\tau$≥st1, the first gradient starts with the following profile:

$$\frac{C_{f,Ns}(\tau)}{C_{0,Ns}} = a1 + a2(\tau - \tau_{imp}) + a3(\tau - \tau_{imp})^2 + a4(\tau - \tau_{imp})^3$$
$$+ a5\exp\left(a6(\tau - \tau_{imp})\right) \ . \tag{9-4}$$

This means that, at $\tau$=st1, the gradient front hits the column inlet. When $\tau$>st2, the second gradient starts with the following profile:

$$\frac{C_{f,Ns}(\tau)}{C_{0,Ns}} = a11 + a22(\tau - \tau_{imp}) + a33(\tau - \tau_{imp})^2 + a44(\tau - \tau_{imp})^3$$
$$+ a55 \exp[a66(\tau - \tau_{imp})] \quad . \tag{9-5}$$

If the user wants a different gradient scheme including stepwise gradients, the code can easily be modified in the respective places.

In the two gradient profiles above, $C_{f,Ns}(\tau)$ is the modulator concentration at the column inlet at dimensionless time $\tau$. $\tau$=0 is the moment when the sample first contacts the column inlet. Ideally, the gradient front immediately follows the end of the injected sample stream as shown in Eq. (9-3), thus we have ($\tau - \tau_{imp}$) instead of $\tau$ in the gradient profiles above. In reality, the sample leaves the sample loop at $t$=0, and it enters the column soon after. The delay time between the injector exit and the column inlet is usually negligible. On the other hand, the gradient front begins at the gradient mixer at $t$=0, not at the end of the sample stream especially when the sample loop is not full. The gradient front has to travel through the sample loop to reach the column inlet. There is usually a distance between the gradient front and the end of the sample stream. This creates a time lag during which the modulator concentration is still at its initial concentration, cnspo. A failure to consider this factor may lead to a wrong gradient profile input. Let us consider a simple case in which a gradient elution has a single linear gradient with a dimensionless modulator concentration change of $\Delta c_{Ns}$ (i.e., the gradient starts from cnspo to cnsp0+$\Delta c_{Ns}$) in a dimensionless time period of $\Delta\tau$. The dimensionless sample size is $\tau_{imp}$. At time $\tau$=0, the sample starts to enter the column inlet and the gradient front starts to leave the gradient mixer. Assume that it takes the gradient front a dimensionless time of $\tau_{delay}$ to travel from the mixer to the column inlet via the sample loop. Then we should have st1=$\tau_{delay}$, and the gradient profile should be

$$\frac{C_{f,Ns}(\tau)}{C_{0,Ns}} = \left[ cnsp0 - \frac{\Delta c_{Ns}}{\Delta\tau}(\tau_{delay} - \tau_{imp}) \right] + \frac{\Delta c_{Ns}}{\Delta\tau}(\tau - \tau_{imp}) \tag{9-6}$$

which means that in the input data file, $a1$=cnsp0-($\Delta c_{Ns}/\Delta\tau$)($\tau_{delay}-\tau_{imp}$), and $a2$=($\Delta c_{Ns}/\Delta\tau$). $a2$ is obviously the gradient slope. If $\tau_{delay}$=$\tau_{imp}$, meaning the gradient front immediately follows the end of injected sample stream, then st1=$\tau_{delay}$=$\tau_{imp}$, and Eq. (9-6) simplifies to

$$\frac{C_{f,Ns}(\tau)}{C_{0,Ns}} = cnsp0 + \frac{\Delta c_{Ns}}{\Delta\tau}(\tau - \tau_{imp}) \quad . \tag{9-7}$$

Thus, $a1$=cnsp0 in the input data file.

In Fig. 9.2, the following quadratic gradient is used to replace the two linear gradients in Fig. 9.1 to achieve similar baseline separation,

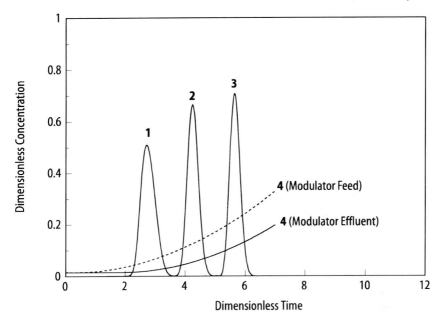

**Fig. 9.2.** Gradient elution with a nonlinear gradient

$$\frac{C_{f,Ns}(\tau)}{C_{0,Ns}} = 0.015 + 0.007(\tau - \tau_{imp})^2 \quad \text{for } \tau \geq \tau_{imp} \quad .$$

In the input file, the st2 value is set to 10 that is larger than tmax=7. This signals the code that there is only one gradient involved in the simulation, because the code will automatically stop calculation when it reaches tmax=7.

## 9.5  Summary

In this chapter, a general rate model has been presented for the study of gradient elution in nonlinear chromatography. The model is suitable for preparative- and large-scale chromatography since various dispersive effects, such as axial dispersion, film mass transfer, and intraparticle diffusion are considered. The eluite-modulator relationship in the model accounts for both electrostatic and hydrophobic interactions. The Fortran code based on the model provides a useful tool for studying various aspects of gradient elution chromatography.

# 10 Multicomponent Radial Flow Chromatography

## 10.1 Introduction

Radial flow chromatography (RFC) was introduced into the commercial market in the mid-1980s [100] as an alternative to the conventional axial flow chromatography (AFC) for preparative- and large-scale applications. Compared to AFC, the RFC geometry (Fig. 10.1) provides a relatively large flow area and a short flow path. It allows a higher volumetric flow rate with a lower bed pressure compared to longer AFC columns. If soft gels or affinity

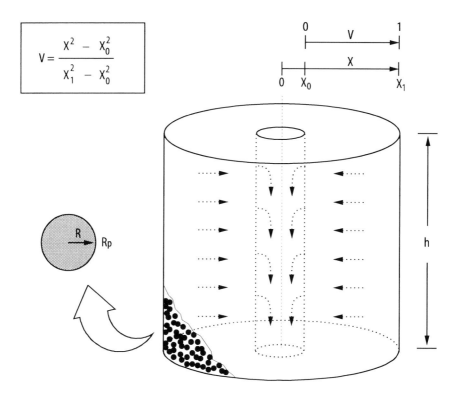

$$V = \frac{X^2 - X_0^2}{X_1^2 - X_0^2}$$

**Fig. 10.1.** Radial Flow Chromatography (RFC) column showing inward flow

matrix materials are used as separation media, the low pressure drop of RFC helps prevent bed compression [21, 101]. RFC columns, both prepacked and unpacked, with a range of size from 50 milliliters to 200 liters in bed volume are commercially available. An experimental case study of the comparison of RFC and AFC was carried out by Saxena and Weil [102] for the separation of ascites using QAE cellulose packings. They reported that by using a higher flow rate, the separation time for RFC was a quarter of that needed for a longer AFC column with the same bed volume. It was claimed that by using RFC instead of AFC, separation productivity can be improved quite significantly [100].

Mathematical modeling of RFC presents certain challenges. Since the linear flow velocity ($v$) in the RFC column changes continuously along the radial coordinate of the column, the radial dispersion and external mass transfer coefficients may no longer be considered constant. This important feature was rarely considered in the mathematical modeling of RFC in the literature in the past. Extensive theoretical studies have been reported for single component ideal RFC, which neglects radial dispersion, intraparticle diffusion, and external mass transfer resistance. In such studies, a local equilibrium assumption and linear isotherms are often assumed. The earliest theoretical treatment of RFC was made by Lapidus and Amundson [103]. A similar study was carried out by Rachinskii [104]. Later, Inchin and Rachinskii [105] included bulk-fluid phase molecular diffusion in their modeling. Lee et al. [30] proposed a unified approach for moments in chromatography, both AFC and RFC. They used severa RFC l single component rate models for the comparison of statistical moments for RFC and AFC. Their models included radial dispersion, intraparticle diffusion, and external mass transfer effects. Kalinichev and Zolotarev [106] also carried out an analytical study on moments for single component RFC in which they treated the radial dispersion coefficient as a variable.

A rate model for nonlinear single component RFC was solved numerically by Lee [31] by using the finite difference and orthogonal collocation methods. His model considered radial dispersion, intraparticle diffusion, external mass transfer, and nonlinear isotherms. It used averaged radial dispersion and mass transfer coefficients instead of treating them as variables. A nonlinear model of this kind of complexity has no analytical solution and must be solved numerically.

Rhee and Amundson [9] discussed the extension of their multicomponent chromatography theory for ideal AFC with Langmuir isotherms to RFC. Apart from this, so far no other detailed theoretical treatment of nonlinear multicomponent RFC is available in the literature.

With the development of powerful computers and efficient numerical methods, more complicated treatment of multicomponent RFC now becomes possible. A general model for multicomponent RFC can provide some useful information. In this chapter, a numerical procedure is presented for solution to a general rate model for multicomponent RFC. The model is solved by using the same basic approached presented in Chap. 3 for AFC models. The solution of the model enables the discussion of several important issues concerning the characteristics and performance of RFC and its differences from AFC, and the question of whether one should treat dispersion and mass transfer coefficients as variables.

## 10.2 General Multicomponent Rate Model for RFC

Figure 10.1 shows the structure of a cylindrical RFC column. The following basic assumptions are needed for the formulation of a unified general model for RFC.

(1) The chromatographic process is isothermal. There is no temperature change during a run.
(2) The porous particles in the bed are spherical and uniform in diameter.
(3) The concentration gradients in the axial direction are negligible. This means that the maldistribution of radial flow is ignored.
(4) The fluid inside particle macropores is stagnant, i.e., there is no convective flow inside macropores.
(5) An instantaneous local equilibrium exists between the macropore surfaces and the stagnant fluid in the macropores.
(6) The film mass transfer theory can be used to describe the interfacial mass transfer between the bulk-fluid and particle phases.
(7) The diffusional and mass transfer coefficients are constant and independent of the mixing effects of the components involved.

Based on these basic assumptions, Eqs. (10-1) and (10-2) are formulated from the differential mass balance for each component in the bulk-fluid and particle phases. In Eq. (10-1), the "+" sign represents outward flow, and "-$v$" inward flow.

$$-\frac{1}{X}\frac{\partial}{\partial X}\left(D_{bi}X\frac{\partial C_{bi}}{\partial X}\right)\pm v\frac{\partial C_{bi}}{\partial X}+\frac{\partial C_{bi}}{\partial t}+\frac{3k_i(1-\varepsilon_b)}{\varepsilon_b R_p}\left(C_{bi}-C_{pi,R=R_p}\right)=0 \quad (10\text{-}1)$$

$$(1-\varepsilon_p)\frac{\partial C_{pi}^*}{\partial t}+\varepsilon_p\frac{\partial C_{pi}}{\partial t}-\varepsilon_p D_{pi}\left[\frac{1}{R^2}\frac{\partial}{\partial R}\left(R^2\frac{\partial C_{pi}}{\partial R}\right)\right]=0 \quad . \qquad (10\text{-}2)$$

The initial conditions for the PDE system are:

at $t=0$,   $C_{bi}=C_{bi}(0,X)$   and   $C_{pi}=C_{pi}(0,R,X)$   . $\qquad\qquad$ (10-3,4)

The boundary conditions are:

at the inlet $X$ position, $\qquad \dfrac{\partial C_{bi}}{\partial X}=\dfrac{v}{D_{bi}}\left(C_{bi}-C_{fi}(t)\right)$ $\qquad\qquad$ (10-5) and at the outlet $X$ position,    $\dfrac{\partial C_{bi}}{\partial X} = 0$  .    (10-6)

Equations (10-1) and (10-2) can be written in dimensionless forms as follows.

$$-\frac{\partial}{\partial V}\left(\frac{\alpha}{Pe_i}\frac{\partial c_{bi}}{\partial V}\right) \pm \frac{\partial c_{bi}}{\partial V} + \frac{\partial c_{bi}}{\partial \tau} + \xi_i\left(c_{bi} - c_{pi,r=1}\right) = 0 \tag{10-7}$$

$$\frac{\partial}{\partial \tau}\left[\left(1-\varepsilon_p\right)c_{pi}^* + \varepsilon_p c_{pi}\right] - \eta_i\left[\frac{1}{r^2}\frac{\partial}{\partial r}\left(r^2\frac{\partial c_{pi}}{\partial r}\right)\right] = 0 \tag{10-8}$$

In Eq. (10-7), the dimensionless variable, $V = (X^2 - X_0^2)/(X_1^2 - X_0^2) \in [0,1]$, is based on the local volume averaging method [31]. $\alpha = 2\sqrt{V + V_0}$ $\times\left(\sqrt{1+V_0} - \sqrt{V_0}\right)$ is a function of $V$.
The initial conditions are:

at $\tau = 0$,   $c_{bi} = c_{bi}(0,V)$,   and   $c_{pi} = c_{pi}(0,r,V)$  .    (10-9,10)

The boundary conditions are:

$$\frac{\partial c_{bi}}{\partial V} = Pe_i\left[c_{bi} - \frac{C_{fi}(\tau)}{C_{0i}}\right] \quad . \tag{10-11}$$

At the inlet $V$ position:

For frontal adsorption,          $C_{fi}(\tau)/C_{0i} = 1$  .

For elution,          $C_{fi}(\tau)/C_{0i} = \begin{cases} 1 & 0 \leq \tau \leq \tau_{imp} \\ 0 & \text{otherwise} \end{cases}$  .

After the introduction of a sample in the form of a rectangular pulse:

If component $i$ is displaced,          $C_{fi}(\tau)/C_{0i} = 0$

If component $i$ is a displacer,          $C_{fi}(\tau)/C_{0i} = 1$  .

At the outlet $V$ position,          $\partial c_{bi}/\partial V = 0$  .

For the particle phase governing equation, the boundary conditions are:

at $r = 0$, $\partial c_{pi}/\partial r = 0$; and at $r = 1$, $\partial c_{pi}/\partial r = Bi_i(c_{bi} - c_{pi,r=1})$  .   (10-12,13)

Note that all the dimensionless concentrations are based on $C_{0i}$, that is the maximum of the feed profile $C_{fi}(\tau)$ for each component.

For RFC the radial dispersion coefficient $D_{bi}$ is a variable that depends on the linear velocity $v$. In liquid chromatography it can be assumed [31, 65, 67] that $D_{bi} \propto v$. Thus $Pe_i = v(X_1 - X_0)/D_{bi}$ can be considered constant in liquid RFC. The variation of $Bi_i$ values observes the following relationship:

$$Bi_i \propto k_i \propto v^{1/3} \propto \left(\frac{1}{X}\right)^{1/3} \propto \left(V + V_0\right)^{-1/6} \quad . \tag{10-14}$$

If $Bi_{i,V=1}$ values are known, $Bi_i$ values anywhere else can be obtained from Eq. (10-15).

$$Bi_{i,V} = \left(\frac{1+V_0}{V+V_0}\right)^{1/6} Bi_{i,V=1} \quad .$$

(10-15)

$\xi_i$ can be calculated from $Bi_i$ using its definition $\xi_i = 3Bi_i\eta_i(1-\varepsilon_b)/\varepsilon_b$.

## 10.3 Numerical Solution

The strategy for the numerical solution to the model is identical to that used for the general rate model in Chap. 3. The PDE system of the governing equations is first discretized. The finite element and orthogonal collocation methods are used to discretize the bulk-fluid phase and the particle phase governing equations, respectively. The resulting ordinary differential equation (ODE) system is then solved using the IMSL subroutine "IVPAG".

In the numerical procedure, $D_{bi}$ and $k_i$ values are treated as variables that are dependent on the variation of $v$ along the radial coordinate $V$. Meanwhile, intraparticle diffusivities ($D_{pi}$) are regarded as independent of the variations of $v$. Due to the special geometry of RFC, there are two space coordinate ($V$) dependent variables, $\alpha$, and $\xi_i$. The finite element method can deal with this situation routinely without any extra trouble. The ability to deal with variable physical properties with ease is one of the advantages of the finite element method. Accuracy is another notable advantage of the method.

The accommodation of variable $Bi_i$ in the particle phase is also easy. Since particle phase equations must be solved at each finite element node (with given nodal position, $V$) in the function subroutine of the Fortran 77 code, $Bi_{i,V}$ values can be readily obtained from Eq. (10-15). The Fortran 77 code for the simulation of RFC is named "RATERFC.F." The code is similar to the Fortran 77 code RATE.F for AFC presented in Chap. 3.

A study of the effects of treating $D_{bi}$ and $k_i$ as variables compared to treating them as constants has been carried out by Gu et al. [65]. The comparison between RFC and AFC was also studied through computer simulation.

## 10.4 How to Use the Fortran 77 Code RATERFC.F

To demonstrate the usage of the computer code, the multicomponent Langmuir isotherm is used for simulations. Parameter values used in the simulations are listed in Table 10.1, or mentioned during discussions. Figure 10.2 shows simulated breakthrough curves for two components in inward flow RFC.

**Table 10.1.** Parameter values used for simulation in Chap. 10[*]

| Figure(s) | Species | Physical Parameters | | | | | Numerical Parameters | |
|---|---|---|---|---|---|---|---|---|
| | | $Pe_{Li}$ | $\eta_i$ | $Bi_i$ | $a_i$ | $b_i \times C_{0i}$ | Ne | N |
| 10.2 | 1 | 100 | 10 | 10 | 1 | $2 \times 0.2$ | 15 | 2 |
| | 2 | 80 | 8 | 8 | 10 | $20 \times 0.2$ | | |
| 10.3 | 1 | 100 | 10 | 10 | 1 | $2 \times 0.2$ | 13 | 2 |
| | 2 | 120 | 12 | 12 | 10 | $20 \times 0.2$ | | |
| 10.4 | 1 | 100 | 1 | 40 | | | 6 | 2 |
| | 2 | 100 | 1 | 40 | | | | |
| | 3 | 100 | 1 | 40 | | | | |

[a] In all runs, $\varepsilon_p=0.4$ and $\varepsilon_b=0.4$, and $V_0=0.04$. In Figs. 10.2 and 10.3, the sample size is $\tau_{imp}=0.2$. The $c^\infty$, $C_{0i}/C_{01}$, $Da_i^a$, $Da_i^d$ and $F_i^{ex}$ values for Fig. 10.4 are the same as those for Fig. 8.12 in Chap. 8, which are listed in Table 8.1.

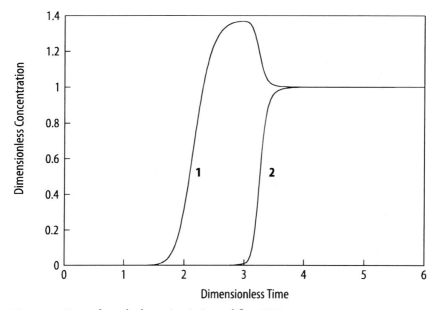

**Fig. 10.2.** Binary frontal adsorption in inward flow RFC

The Fortran 77 code RATERFC.F requires a data file named "data" to provide simulation parameters upon execution of the code. Below is the heading generated by the code when it is executed for Fig. 10.2.

```
Radial Flow Chromatogr. Simulator by T. GU (Ohio U.)
====================================================================
in/outward (-1/1)  VO           iave (1,2/0 for y,y/n)
        -1         0.04000          0
_____

nsp nelemb nc index  timp   tint  tmax  epsip epsib
 2    15    2   1    0.500  0.015  4.0   0.400 0.400
_____
```

```
   PeL         eta          Bi        CO      consta    constb
 100.00     10.000     10.000    0.20000     1.000     2.000
  80.00      8.000      8.000    0.20000    10.000    20.000
===================================================== End of data file.
Total ODE =  186     Data pts =   266
tol  =    1.0000000000000D-05   <- Double precision or not, see this.
```

```
index =1 Breakthru; =2, Elution with inert MP
index =3 Step-change disp. Last comp. is displacer
index =4 BT, switch to displacement at t = tshift
index =5 Same as index=4, but reverse flow
index =6 Elution, the last component is modifier
index =7 Same as =6, but sample is in inert...
index =10+ use separation factors
Input Bi value should be at its value at V=1.
If iave=0, then the code treats k, D as variables
If iave=1,2 an ave Bi is used by the code from Bi|V=1.
If iave=1, taking ave at (X1+X0)/2
If iave=2, taking ave at V=0.5
If spiral flow chromatography, in/outward (-10/10)
```

```
 Results (t, c1, c2,...) follow.    Please wait...

   0.01500    0.00000    0.00000
   (... more data points)
```

Stripping away the text lines in the heading above from the beginning to the "End of data file", the remaining numerical figures are what the file "data" contains. The first three inputs are the flag number for inward flow (input value =-1) or outward flow RFC (input value=1), the value for $V_0$ ($V0$), and the flag number for how to treat $D_{bi}$ and $k_i$ (iave). If iave=0, the code treats $D_{bi}$ and $k_i$ as variables. In order to study the effect of treating $D_{bi}$ and $k_i$ as variables compared to taking an averaged value [65], the code allows the user to take averaged $D_{bi}$ and $k_i$ values at $(X_1+X_0)/2$, or at $V=0.5$ for the calculation of $Pe_i$, $Bi_i$, and $\xi_i$ by setting iave=1 or 2, respectively. Other input data for the code are identical to those used for RATE.F. The user may consult Chap. 3 for details.

Fig. 10.3 shows a binary elution in inward flow RFC. The code can also simulate other operations such as step-change displacement, etc. More simulated chromatograms are given elsewhere by Gu et al. [65].

## 10.5 Extensions of the General Multicomponent Rate Model for RFC

Extensions to the basic general rate model for AFC in Chap. 3 have also been applied to the RFC model. These include second order kinetics, size exclusion effect and a reaction in the liquid phase for the modeling of biospecific elution using a soluble ligand. The extensions have been carried out with

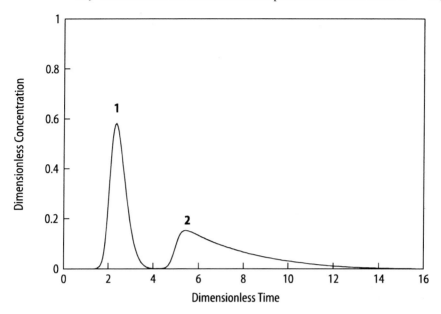

**Fig. 10.3.** Binary elution with an inert mobile phase in inward flow RFC

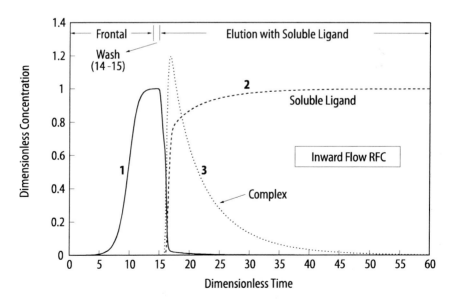

**Fig. 10.4.** Simulation of affinity RFC with inward flow

ease. Details are omitted here since the necessary modifications for adding second order kinetics involve only the particle phase governing equation, in which the AFC and the RFC model do not differ, except that the $k_i$ values in RFC are variables.

The adding of reaction terms for the interaction between a macromolecule and the soluble ligand involves the bulk-fluid phase, but it does not touch the characteristic terms of the RFC model. It has been easily implemented. Figure 10.4 shows a simulated chromatogram of an affinity separation process with frontal adsorption, wash, and elution stages. The Fortran 77 code used for the simulation is named "AFFRFC.F".

## 10.6 Summary

A general rate model for RFC has been presented. Radial dispersion and mass transfer coefficients are treated as variables in the model. The model is solved numerically by using the finite element and orthogonal collocation methods for the discretization of bulk-fluid and particle phase PDEs, respectively. Several chromatographic operations have been simulated.

# 11 Scale-Up of Liquid Chromatography using General Rate Models

One of the critical factors for a successful scale-up of liquid chromatography using the rates models is accurate parameter estimations. Three types of parameter are needed to carry out model calculations using the rate models. Isotherm parameters, the particle porosity and the bed void volume fraction are important to the accuracy of model calculations. Physical dimensions of the column are equally important, but they can be easily and precisely measured. Less important parameters are the mass transfer coefficients, which usually do not affect the general location of an elution peak. They affect the shape of the peak. However, such an influence is not very sensitive to the fluctuation of the mass transfer coefficients. Thus, parameter estimation of these coefficients does not have to be very stringent.

## 11.1 Isotherms

Isotherms, especially multicomponent isotherms are generally not available for a particular system from existing publications in the literature. They may have to be measured experimentally. Isotherm data points are first measured experimentally and then fitted or correlated with an isotherm model, such as the most commonly used Langmuir isotherm model. For a single component system, the isotherm data can always be correlated using a polynomial equation with reasonable accuracy if there are no other good isotherm models to use. Unfortunately, the simple polynomial model is not a substitute for a multicomponent isotherm model. There are generally two experimental methods for obtaining isotherm data.

### 11.1.1 Batch Adsorption Equilibrium Method

This is a straightforward method. A solution with a fixed concentration of solute is mixed with a certain amount of adsorbent in a test tube. The content is then well stirred or shaken for a sufficient amount of time until an adsorption equilibrium is established. The supernatant is then analyzed for the liquid phase concentration. Suppose a solution with $m_0$ moles of a solute is mixed with $W$ g of an adsorbent that has a post-swelling density of $\rho_p$ (grams of particle per unit volume of adsorbent particle skeleton). The equilibrium concentration in the stationary phase can be obtained using the following relationship:

$$C_p^* = \frac{\left[ m_0 - C_p \left( V_m - \dfrac{W}{\rho_p} \right) \right] \rho_p}{W} \tag{11-1}$$

where $C_p$ is the equilibrium concentration of the solute in the supernatant, and $V_m$ is the total volume of the solution mixed with the adsorbent. With a set of experimental data points of $C_p^*$ vs $C_p$, an analytical expression of a single component isotherm can be obtained by correlating these data points using an isotherm model, such as the Langmuir isotherm.

### 11.1.2 Column Method

If the adsorbent is not available or is too expensive for the batch method, the frontal adsorption method [107] can be used instead to obtain isotherms. In this method, a mini-column packed with the adsorbent is used for frontal adsorptions with step-changes of solute concentration in the mobile phase. The column response is then analyzed to yield the adsorbed solute amount. The isotherm data can then be calculated based on the amount of the adsorbent packed inside the column. The advantage of this method is that it saves the adsorbent. Jacobson et al. [107] have provided a detailed experimental procedure and equations for the evaluation of isotherm data. It must be pointed out that different researchers may have different units for the particle phase concentration, $C_p^*$. Before an isotherm can be used for the rate models in this book, the unit of $C_p^*$ must be converted to moles of solute adsorbed per unit volume of particle skeleton.

### 11.1.3 Langmuir Isotherm

The Langmuir isotherm (Eq. 3-17) is most commonly used in liquid chromatography. Let us first discuss the measurement of a single component Langmuir isotherm. The standard batch adsorption method or the column method can be used for the experiments. The following method is a variation of the column method, in which one does not have to rely on getting a clear-cut frontal breakthrough curve, and the solute solution does not always have to be pure.

A mini-column is saturated with a solution which has a solute concentration of $C_0$ using a frontal breakthrough. The amount of solute remaining in the column ($m_1$ mol) can be determined by calculating the difference between the total amount of solute pumped into the column and the total amount in the flow-through. The equilibrium concentration in the particle phase $C_p^*$ can be calculated using the following equation:

$$C_p^* = \frac{m_1 - C_0 V_b \left[ \varepsilon_b + (1 - \varepsilon_b) \varepsilon_p \right]}{V_b (1 - \varepsilon_b)(1 - \varepsilon_p)} \qquad (11\text{-}2)$$

where $V_b$, $\varepsilon_b$, $\varepsilon_p$ are the bed volume, the bed void volume fraction, and the particle porosity, respectively. $\varepsilon_b$ and $\varepsilon_p$ values may be available from the vendor. They can also be measured. With the experimental dead volume time of unretained large molecules (such as blue dextrin), $t_d$, one can obtain $\varepsilon_b$ from the following relationship,

$$t_d = \frac{L}{v} = \frac{\pi d^2 L \varepsilon_b}{4Q} \qquad (11\text{-}3)$$

where $L$, $v$, $d$, $Q$ are the column length, the interstitial velocity of the mobile phase, the column inner diameter, and the mobile phase volumetric flow rate, respectively. By measuring $t_0$, the dead volume time of unretained small molecules such as salt and solvent molecules that penetrate the particle macropores, one can evaluate $\varepsilon_p$ from the following relationship:

$$t_d = \frac{t_0}{\dfrac{\varepsilon_b + (1 - \varepsilon_b) \varepsilon_p}{\varepsilon_b}} \qquad . \qquad (11\text{-}4)$$

$t_0$ is also called "the solvent time" by some chromatographers. It is larger than $t_d$, since small molecules penetrate the macropores. Equations (11-3) and (11-4) are quite easy to obtain if one understands that $t_d$ and $t_0$ in elutions have the same values as the breakthrough times in single component frontal analyses for a large unretained solute and a small unretained solute, respectively. Equations (11-3) and (11-4) can be readily derived based a mass balance using the concept of column holdup capacity.

The Langmuir isotherm for the single component frontal adsorption experiment gives

$$C_p^* = \frac{b C^\infty C_0}{1 + b C_0} \qquad (11\text{-}5)$$

in which the $b$ value is still unknown. An isocratic elution with a dilute sample on the column gives the retention factor $k'$ (also known as the capacity factor, which is defined as the ratio of the total moles of solute in the stationary phase to that in the mobile phase at equilibrium) based on the following relationship by Snyder et al. [2]:

$$k' = \frac{t_R - t_0}{t_0} \qquad (11\text{-}6)$$

where $t_R$ is the retention time of the solute. Based on the definition of capacity factor, it is easy to show that, for an isocratic elution with a dilute sample which observes the linear range of the Langmuir isotherm,

$$k' = \phi C^{\infty} b \qquad (11\text{-}7)$$

where $\phi$ is the phase ratio (the stationary phase particle skeleton volume to mobile phase volume including the particle macropores). $\phi$ can be determined from the following equation based on its definition

$$\phi = \frac{(1 - \varepsilon_b)(1 - \varepsilon_p)}{\varepsilon_b + (1 - \varepsilon_b)\varepsilon_p} \qquad (11\text{-}8)$$

Solving Eqs. (11-5) and (11-7) simultaneously gives the $b$ and $C^{\infty}$ values for the solute and the packing inside the column at the temperature that is used to carry out the two experiments. If a modulator (modifier) is involved in the chromatography, the frontal adsorption and the isocratic elution experiments must be carried out using the same modulator concentration, otherwise Eqs. (11-5) and (11-7) cannot be coupled together, since the $b$ value depends on the modulator concentration.

If we assume that a multicomponent system follows the Langmuir isotherm (Eq. 3-19) with a uniform saturation capacity for all the components, parameters $a_i(=b_i C^{\infty})$ and $b_i$ in the isotherm will be the same as those in their corresponding single component isotherms. If the $C^{\infty}$ value has already been determined using one of the components, any remaining $b_i$ values may be determined using a single isocratic elution with a dilute sample containing all the remaining components to yield their retention times. The $b_i$ values are then calculated using Eqs. (11-6) and (11-7).

It is helpful to point out that, in all isocratic elution experiments described above, the sample does not have to be pure since dilute samples do not have any interference between different components. The individual retention times are not affected. But in the frontal breakthrough experiment the sample has to be pure. If a pure sample is not available, one may use a substitute pure compound to obtain a set of b and $C^{\infty}$ values. This $C^{\infty}$ (mole solute per unit volume of particle skeleton) value may be used for the multicomponent system if the substitute compound is very similar to the first compounds.

Snyder et al. [107] developed a method to measure the column saturation capacity based on the slightly different retention times of two gradient elutions, one using a small sample size, the other using a large overload sample size. They provided a semi-empirical equation for the calculation of the column saturation capacity, which can be converted to get $C^{\infty}$.

For multicomponent gradient elutions described in Chap. 9, $C^{\infty}$ is considered equal for all the eluites (solutes other than the modulator) independent of the modulator concentration, and the $b_i$ values are evaluated as a function of the modulator concentration. One may use the modified column method above or the method by Snyder et al. to obtain the $C^\infty$ value using a fixed modulator concentration. This usually requires several isocratic runs using different modulator concentrations. With these isotherm data points for a component, we can evaluate the parameters in the eluite-modulator relationship such as Eq. (9-1), which is suitable for hydrophobic interaction chromatography. Equation (9-1) may be used for reversed-phase chromatography if there is no better correlations. As a matter of fact, one can pick any eluite-modulator relationship as long as it fits the experimental data well. In the measurement of the $b_i$ values as a function of modulator concentration, a very small mini-column is recommended to carry out the isocratic elutions since the retention times corresponding to different modulator concentrations can be orders of magnitude different. Using a very small mini-column helps cut down the retention time and reduce the band spreading [108].

### 11.1.4 Other Isotherm Models

In his book, *Principles of Adsorption and Adsorption Processes*, Ruthven [5] devoted a chapter to the discussion of many different kinds of multicomponent isotherm model apart from the Langmuir isotherm model, such as the Langmuir-Freundlich Equations, the General Statistical Model, the Dubini-Polanyi Theory, the Ideal Adsorbed Solution Theory (IAST), and the Vacancy Solution Theory. These models were first used in gas-solid absorption or adsorption of liquid hydrocarbons on solid adsorbents such as zeolite. Some of the models, such as the IAST, have been used in protein adsorptions on chromatographic media with certain degrees of success.

Recently, Syu et al. [109] used an artificial neural network for multicomponent isotherms. This method is attractive for multicomponent systems that do not follow the multicomponent Langmuir isotherm and other existing isotherm models. The neural network isotherm prediction system can be incorporated into the computer codes for the rate models used in this book.

In some practical multicomponent systems, the interference effect may be negligible. For example, in a binary elution system, if the two components quickly separate from each other, the duration of the interference effect will be short, thus the system may be treated as two separate single component systems. This may happen if the sample size is small or the affinity difference of the two components is large. Discussions provided in Chap. 5 should be helpful in making such a judgement. Several close components may be lumped together and be treated as a pseudo-component.

## 11.2 Mass Transfer Parameters

In this book, it is assumed that mass transfer and diffusional parameters of different components in a multicomponent system are the same as those in a single component system, i.e., there are no mixing effects.

Mass transfer parameters such as $k_i$, $D_{bi}$, $D_{pi}$ are often not available from literature, or not easily measured by experiments. However, they can be estimated with certain accuracy. Fortunately, rate models are not very sensitive to mass transfer parameters. Errors up to a certain degree do not affect the outcome to any great extent.

It should be pointed out that using a general rate model for parameter estimation is not a good practice. Chap. 4 shows that changes in different mass transfer parameters may offset or compensate each other. One must use limiting cases for parameter estimation by using a degenerated rate model, i.e., minimizing other effects before measuring one effect. Otherwise unpredictable large errors may occur.

The following correlation from Wakao et al. [110] may be used to obtain the film mass transfer coefficient $k$,

$$Sh = 2.0 + 1.45 Sc^{1/3} Re^{0.5}    (Re < 100) \tag{11-9}$$

where $Sh = k(2R_p)/D_m$, $Sc = \mu/(\rho_f D_m)$, and $Re = v\rho_f(2R_p)/\mu$. $D_m$ is the molecular diffusivity, and $\rho_f$ is the fluid density."

If the axial dispersion coefficient $D_{bi}$ values are not easily available, Peclet numbers may be directly estimated using the following experimental correlation obtained by Chung and Wen [111] for fixed-beds:

$$Pe_L = \frac{L}{2R_p \varepsilon_b}(0.2 + 0.011 Re^{0.48}) \quad . \tag{11-10}$$

A simpler, maybe more rough estimation, can be obtained by using the relationship derived from Eq. (4-1) in Chap. 4, which states $D_b = \gamma_2(2R_p)v$. Thus, $Pe_L = vL/D_b = 1/(2\gamma_2 R_p)$ where $\gamma_2$ may be given a value of 0.5 m$^{-1}$ according to Ruthven [5].

The intraparticle diffusivity $D_{pi}$ can be obtained from a correlation used by Yau et al. [55]:

$$D_p = D_m(1 - 2.104\lambda + 2.09\lambda^3 - 0.95\lambda^5)/\tau_{tor} \tag{11-11}$$

in which $\lambda = d/d_p$. The tortuosity factor $\tau_{tor} = 2.1 \sim 2.4$. $d_p$ is the pore diameter of the particles. $d$ is the molecular diameter of the solute. If we assume spherical molecules, $d$ can be readily obtained from the molecular volume. The determination of molecular volume from molecular weight and specific volume was discussed by Cantor and Schimmel [112]. For unhydrated spherical molecules, it is easy to show that $d(\text{Å}) = 1.47(MW \cdot V_s)^{1/3}$ where $V_s$ is the specific volume (ml/g).

Truei et al. [113] used some of the correlations above for parameter estimations in a case study involving gradient elution of four proteins in hydrophobic interaction chromatography. A satisfactory agreement was achieved between model calculations and experimental results. If a particular mass transfer parameter cannot be evaluated, a typical value may be used as a substitute. As a matter of fact, if one is not very interested in exact band widths of peaks, a set of values can be artificially assigned to the three mass transfer related dimensionless groups, Pe$_i$, η$_i$ and Bi$_i$, to carry out an initial simulation.

## 11.3  Evaluations of Pe$_{Li}$, η$_i$, and Bi$_i$

The values of Pe$_i$, η$_i$, and Bi$_i$ are calculated based on their definitions. They require mass transfer parameters, physical measurements of the particles and the bed. In some cases, a set of these values can be assigned to a particular compound without parameter estimations. The most important example is the solvent or the salt used for gradient elution chromatography. The solvent or the salt is typically a small molecular and it migrates inside the column with little dispersion. This is also due to the fact that the solvent or salt solution is often pumped into the column continuously unlike a pulse injection. In addition, the salt or solvent is often not retained at all. The actual measured values of Pe$_i$, η$_i$, and Bi$_i$ can be very large, making the model system extremely stiff. This situation can be avoided by artificially choosing smaller values for the three dimensionless parameters. The elution profile of the solvent or salt may change slightly, and the difference can easy be verified by running simulation with blank gradients and comparing the elution profiles. Usually if we assign 500, 5 and 5 to Pe$_i$, η$_i$, and Bi$_i$ for a solvent or salt, respectively, the result should be little different from that calculated with actual estimated parameters.

Generally speaking, in a practical simulation, if a component makes the model system extremely stiff such that the simulation requires a lot of time, or even fails to convergence, one may reduce the Pe$_i$, η$_i$ values for that component to make the system less stiff. This may have little or no effect on other components (such as proteins) that are likely of more importance. The modified result may reduce the peak height for that stiff component, but its retention time is not affected and its band width will not change much. A peak's retention time and band width are most important during scale-up. Of course, if it is manageable, one should always use the actual parameters to run the simulation in order to enhance the realism of the modeling.

## 11.4  General Procedure for Scale-Up

After the isotherm and mass transfer parameters have been estimated, a small column may be used to check whether the model can predict the elution profile of a small sample on that column. If the agreement between the simulated results and the experimental chromatograms is good, then the model may be used to predict the concentration profiles for a large column by changing parameters related to column dimensions, which affect the $Pe_i$, $\eta_i$, and $Bi_i$ values in the data input. If the agreement is not good, it probably means that the isotherm or mass transfer parameters used for the model are not accurate enough. Without further improvement in the accuracy of the parameters, the model should not be used for scale-up.

During a scale-up, if the fitting of concentration profiles is not good, some observations may be made to help find what went wrong. For example, if there is a big discrepancy in retention times between the simulated data and the experimental data, it becomes quite clear that the isotherm used is not accurate, or the bed void volume fraction or particle porosity is inaccurate. More accurate mass transfer parameters will not improve the situation. On the other hand, if the peak tail is more diffused than that shown by simulation, it is likely that the mass transfer rates used in the simulation are too high, assuming that isotherm data are fairly accurate. Remember, concentrations in the nonlinear range of the Langmuir isotherm may also cause sharpening of the peak front and diffusion of the peak tail.

In reality, when a preparative- or large-scale HPLC column is used, the column is usually placed near the end of a multi-stage purification process after the product has been pre-purified using various other techniques, such as ultrafiltration, salt precipitation, or low pressure size exclusion chromatography. This is because it is not economical to apply a "dirty" sample onto an expensive large HPLC column without pre-purification. The sample for the large HPLC column may have several impurities, possibly with one of them very close to the desired product molecule according to the HPLC result from a small column. In such a case, we may pick only this impurity and the product molecule and treat them as a binary mixture for the modeling. The rest of the impurities are ignored in order to simplify the system. It is also possible to treat several impurities that are very close to each other, and are all near the front or the tail of the product peak, as a single "lumped" impurity. The isotherm of the binary mixture can then be measured using a mini-column with the same packing as the large column and the mass transfer parameters estimated. Several vendors in the U.S. sell different series of stainless steel HPLC columns with exactly the same packing in a series, with a bed volume range of several milliliters to several liters.

The scale-up for size exclusion chromatography is much simpler than for other chromatography, since no isotherms are needed. Only one experimental run on a small column with a mixture sample is required. For example, suppose we want to predict the elution profiles of a binary protein mixture on a large size exclusion column. A small column with the same packing is used to run a sample containing blue dextrin, protein 1, protein 2, and a salt to yield retention times $t_d$, $t_{R1}$ $t_{R2}$, $t_0$, respectively. The proteins do not have to be pure. Solving Eqs. (11-3) and (11-4) with $t_d$ and $t_0$ values gives $\varepsilon_b$ and $\varepsilon_p$ values. The Fortran 77 code KINETIC.F can then be used to find the size exclusion factor for protein 1, $F_1^{ex}$, by guessing its value and compare the computer calculated $\tau_{R1}$ with experimental $\tau_{R1}(=t_{R1}v/L)$ for protein 1. In the input data file for KINETIC.F, one may set nsp=1, nelemb=10, nc=1, index=2, timp=0.1, tint=0.01, tmax=4.0, epsip=$\varepsilon_p$, epsib=$\varepsilon_b$, PeL=500, eta=10, Bi=5, Cinf=0, $C_0$=1, Daa=Dad=0, and exf=guessed $F_1^{ex}$ value. Meanings of these input parameter names are explained in Sect. 3.8.1. The three mass transfer related dimensionless groups (PeL, eta and Bi) are arbitrary as long as they do not affect the simulated retention values. Similar to getting $F_1^{ex}$, $F_2^{ex}$ for protein 2 can be based on the experimental $t_{R2}$ value. The, $F_1^{ex}$ and $F_2^{ex}$ values obtained by trial and error through simulation using experimental data of retention times should be the same for a large column as long as it has the same packing material and bed pressure as the small column. The $\varepsilon_b$, $\varepsilon_p$, $F_1^{ex}$, and $F_2^{ex}$ values from the small column can subsequently be used together with estimated PeL(=Pe$_i$), eta(=$\eta_i$), and Bi(=Bi$_i$) values for the large column for computer simulation to predict the large column's effluent histories for the two proteins.

When comparing experimental chromatograms with the simulated chromatograms, it is not necessary to obtain the actual concentrations at different positions of a peak experimentally, which requires the fractionation and analysis of the concentration of each fraction. One may directly compare the experimental detector responses with simulated dimensionless concentrations, since we are most interested in the retention times and band widths of the peaks.

The dimensionless time ($\tau$) in any simulated chromatograms can be converted to dimensional time using the relationship

$$t = \frac{\tau L}{v} + t_{adj} \qquad (11\text{-}12)$$

where $t_{adj}$ is the time needed for the sample to travel from the exit point of the sample injector to the column inlet in the experimental system, and it is often negligible.

Zheng [114] used GRADIENT.F for the scale-up of a small analytical reversed phase HPLC column to a 20 ml preparative HPLC column with the same packing. Isotherm parameters for human growth hormone and an analog of human growth hormone [115] were obtained using the small column by performing a few simple experiments described earlier. Mass transfer parameters were estimated using existing correlations. Good agreements were achieved between simulated gradient elution profiles and actual experimental results for the preparative column. Gu [116] predicted binary elutions in a 44×500 mm size exclusion column using the Fortran code KINETIC.F. Simulated results were in excellent agreement with experimental results.

Using rate models for the scale-up of liquid chromatography holds great potential in future practice. Hopefully, more and more people will realize their values in choosing or custom-make a suitable large column for production from the data obtained on a small column such as an analytical column. Readers may request any of the Fortran programs described in this book by writing to the author, or by sending an electronic mail to guting@ent. ohiou.edu through the Internet. The author welcomes any comments and discussions regarding the use of these programs.

It is helpful to point out that some of the rate models discussed in this book, such as the general rate model with the Langmuir isotherm, and the kinetic model based on second order kinetics can be directly adopted to simulate other fixed-bed processes, such as gas absorption, etc. Models for the simulation of fixed-beds with porous adsorbents have very similar but not identical mass transfer and thermodynamic mechanisms.

# 12 References

1. Gear CW (1972) Numerical Initial-Value Problems in Ordinary Differential Equations. Prentice-Hall, Englewood Cliffs, New Jersey
2. Snyder LR, Glejch JL, Kirkland JJ (1988) Practical HPLC method development. Wiley, New York
3. Levin S, Grushka E (1986) Anal Chem 58:1602
4. Melander WR, El Rassi Z, Horvath Cs (1989) J Chromatogr 469:3
5. Ruthven DM (1984) Principles of adsorption and adsorption processes. Wiley, New York
6. Glueckauf E (1949) Discuss Fraday Soc 7:12
7. Helfferich F, Klein G (1970) Multicomponent chromatography – theory of interference. Marcel Dekker, New York
8. Rhee H-K, Aris R, Amundson NR (1970) Philos Trans R Soc (London) Ser A 267:419
9. Rhee H-K, Amundson NR (1982) AIChE J 28:423
10. Glueckauf E (1947) J Chem Soc 1302
11. Helfferich F, James DB (1970) J Chromatogr 46:1
12. Bailly M, Tondeur D (1981) Chem Eng Sci 36:455
13. Frenz J, Horvath C (1985) AIChE J 31:400
14. Frenz J, Horvath C (1988) High Performance Liquid Chromatography 5:211
15. Martin MJP, Synge RLM (1941) Biochem J 35:1359
16. Yang CM (1980) PhD Thesis, Purdue Univ., Indiana
17. Villermaux J (1981) in: Rodrigues AE, Tondeur D (eds) Percolation processes: Theory and applications. Sijthoff and Noordhoff, Rockville, Maryland
18. Eble J E, Grob RL, Antle PE Snyder LR (1987) J Chromatogr 405:1
19. Seshadri S, Deming SN (1984) Anal Chem 56:1567
20. Solms DJ, Smuts TW, Pretorius V (1971) J Chromatogr Sci 9:600
21. Ernst P (1987) Aust J Biotechnol 1:22
22. Velayudhan A, Ladisch MR (1993) Advances in Biochemical Engineering/Biotechnology 49:123
23. Glueckauf E, Coates JI (1947) J Chem Soc 1315
24. Rhee H-K, Amundson NR (1974) Chem Eng Sci 29:2049
25. Bradley WG, Sweed NH (1975) AIChE Symp Ser 71:59
26. Farooq S, Ruthven DM (1990) Chem Eng Sci 45:107
27. Lin B, Golshan-Shirazi S, Ma Z, Guiochon G (1988) Anal Chem 60:2647
28. Santacesaria E, Morbidelli M, Servida A, Storti G, Carra S (1982) Ind Eng Chem Process Des Dev 21:446
29. Santacesaria E, Morbidelli M, Servida A, Storti G, Carra S (1982) Ind Eng Chem Process Des Dev 21:451
30. Lee W-C, Huang SH, Tsao GT (1988) AIChE J 34:2083
31. Lee W-C (1989) PhD Thesis, Purdue Univ., West Lafayette, IN
32. Chase HA (1984) Chem Eng Sci 39:1099
33. Chase HA (1984) J Chromatogr 279:179
34. Arnold FH, Blanch HW, Wilke CR (1985) J Chromatogr 30:B9
35. Arnold FH, Blanch HW, Wilke CR (1985) J Chromatogr 30:B25

36. Arnold FH, Schofield SA, Blanch HW (1986) J Chromatogr 355:1
37. Arnold FH, Schofield SA, Blanch HW (1986) J Chromatogr 355:13
38. Arve BH, Liapis AI (1987) AIChE J 33:179
39. Arve BH, Liapis AI (1988) Biotech & Bioeng 31:240
40. Gu T, Tsai G-J, Tsao GT (1990) AIChE J 36:784
41. Liapis AI, Rippin DWT (1978) Chem Eng Sci 33:593
42. Mansour A (1989) Sep Sci Technol 24:1047
43. Yu Q, Yang J, Wang N-HL (1989) Comp Chem Eng 13:915
44. Mansour A, von Rosenberg DU, Sylvester ND (1982) AIChE J 28:765
45. Villadsen J, Michelsen ML (1978) Solutions of differential equation models by polynomial approximation. Prentice Hall, Englewood Cliff
46. Finlayson BA (1980) Nonlinear analysis in chemical engineering. McGraw-Hill, New York
47. Antia FD, Horvath C, Ber Bunsenges (1989) Phys Chem 93:961
48. IMSL (1987) IMSL User's Manual, Version 1.0. IMSL, Inc. Houston, Texas, pp. 640-652
49. Pieri G, Piccardi P, Muratori G, Luciano C (1983) La Chimica E L'Industria 65:331
50. Knox JH, Pyper HM (1986) J Chromatogr 363:1
51. Bird RB, Stewart WE, and Lightfoot EN (1960) Transport Phenomena. John Wiley, New York
52. Reddy JN (1984) An Introduction to the Finite Element Method. McGraw Hill, New York
53. Froment GF, Bischoff KB (1979) Chemical reactor analysis and design. Wiley, New York
54. Gu T (1990) Ph.D. Thesis, Purdue Univ., W. Lafayette, Indiana
55. Yau WW, Kirkland JJ, Bly DD (1979) Modern size-exclusion liquid chromatography. Wiley, New York, p. 89
56. Kim DH, Johnson AF (1984) in: Provder T (ed) ACS Symp Series 245:25
57. Koo Y-M, Wankat PC (1988) Separation Science and Technology 23:413
58. Danckwerts PV (1963) Chem Eng Sci 2:1
59. Parulekar SJ (1983) PhD Thesis, Purdue Univ., West Lafayette, IN
60. Parulekar SJ, Ramkrishna D (1984) Chem Eng Sci 39:1571
61. Brian BF III, Zwiebel I, Artigue RS (1987) AIChE Symp Series 83:80
62. Lin B, Ma Z, Guiochon G (1989) Sep Sci Technol 24:809
63. Lee W-C, Tsai G-J, Tsao GT (1990) ACS Symp Series 427:104
64. Lee CK, Yu Q, Kim SU, Wang N-HL (1989) J Chromatogr 484:25
65. Gu T, Tsai G-J, Tsao GT (1991) Chem Eng Sci 46:1279
66. Jonssoon JA (1987) Chromatographic Theory and basic principles. Marcel Dekker, New York
67. Weber SG, Carr PW (1989) in: Brown PR, Hartwick TA (eds) High performance liquid chromatography. Wiley, New York
68. Pfeffer R, Happel J (1964) AIChE J 10:605
69. Wilson EJ, Geankoplis CJ (1966) Ind Eng Chem Fund 5:9
70. Kaizuma H, Myers MN, Giddings JC (1970) J Chromatogr Sci 8:630
71. Horvath C, Lin H-J (1976) J Chromatogr 126:401
72. Giddings JV (1965) Dynamics of Chromatography Part I Principles and Theory. Marcel Dekker, New York, p. 75
73. Yu Q, Yang J, Wang N-HL (1987) Reactive Polymers 6:33
74. Tiselius A (1940) Ark Kem, Mineral Geol 14B:1
75. Wankat PC (1986) Large-scale adsorption and chromatography vol. 1, CRC Press, Cleveland
76. Helfferich F (1962) J Am Chem Soc 84:3242
77. Tiselius A (1943) Ark Kem, Mineral Geol 16A:1
78. Gu T, Tsai G-J, Ladisch MR, Tsao GT (1990) AIChE J 36:1156
79. Gu T, Tsai G-J, Tsao GT (1991) AIChE J 37:1333
80. Thomas WJ, Lombardi JL (1971) Trans Instn Chem Engrs 49:240

81. Hsieh JSC, Turian RM, Tien C (1977) AIChE J 23:263
82. Liapis AI, Litchfield RJ (1980) Chem Eng Sci 35:2366
83. Kapoor A, Yang RT (1987) AIChE J 33:1215
84. Bauman WC, Wheaton RM, Simpson DW (1956) in: Nachod FC, Schubert J (eds) "Ion exclusion" in ion exchange technology. Academic Press, New York
85. Eble JE, Grob RL, Antle PE, Snyder LR (1987) J Chromatogr 384:25
86. Dreux M, Lafosse M, Pequignot M (1982) Chromatographia 15:653
87. Cassidy RM, Fraser M (1984) Chromatographia 18:369
88. Levin S, Grushka E (1987) Anal Chem 59:1157
89. McCormic RM, Karger BL (1980) J Chromatogr 199:259
90. Golshan-Shirazi S, Guiochon G (1988) J Chromatogr 461:1
91. Golshan-Shirazi S, Guiochon G (1989) J Chromatogr 461:19
92. Kemball C, Rideal EK, Guggenheim EA (1948) Trans Faraday Soc 44: 948
93. Liapis AI (1989) J Biotech 11:143
94. Arve BH, Liapis AI (1987) Biotech & Bioeng 30:638
95. Arve BH, Liapis AI (1988) Biotech & Bioeng 32:616
96. Furusaki S, Haruguchi E, Nozawa T (1987) Bioprocess Engineering 2:49
97. Pitts WW Jr (1976) J Chromat Sci 14:396
98. Kang K, McCoy BJ (1989) Biotechnol Bioengng 33:786
99. Gu T, Truei Y-H, Tsai G-J, Tsao GT (1992) Chem Engng Sci 47:253
100. McCormick D (1988) Bio/Technology 6:158
101. Saxena V, Weil AE, Kawahata RT, McGregor WC, Chandler M (1987) Am Lab (Fairfield, CT) 19:112
102. Saxena V, Weil AE (1987) BioChromatography 2:90
103. Lapidus L, Amundson NR (1950) J Phys Colloid Chem 54:821
104. Rachinskii VV (1968) J Chromatogr 33:234
105. Inchin PA, Rachinskii VV (1977) Russ J Phys Chem 47:1331
106. Kalinichev AI, Zolotarev PP (1977) Russ J Phys Chem 51:871
107. Snyder LR, Cox GB, Antle PE (1988) J Chromatogr 3444:303
108. Jacobson J, Frenz J, Horvath C (1984) J Chromatogr 316:53
109. Syu M-J, Tsai G-J, Tsao GT (1993) in Fiechter A (ed) Advances in Biochemical Engineering/ Biotechnology, Vol. 49, Springer, Berlin Heidelberg New York.
110. Wakao N, Oshima T, Yagi S (1958) Kagaku Kagaku 22:780
111. Chung SF, Wen CY (1968) AIChE J 14:857
112. Cantor CR, Schimmel PR (1980) Biophysical Chemistry (Part II). W. H. Freeman, San Francisco, pp. 550-555
113. Truei Y-H, Gu T, Tsai G-J, Tsao GT (1992) in Fiechter A (ed) Advances in Biochemical Engineering/Biotechnology, Vol. 47, Springer, Berlin Heidelberg New York.
114. Zheng Y (1994) PhD Thesis, Ohio Univ., Athens, OH
115. Gu T, Gu Y, Zheng Y, Wiehl PE, Kopchick JJ (1994) Sep Technol 4:258
116. Gu Y (1995) PhD Thesis, Ohio Univ., Athens, OH

# 13 Subject Index

**K. Faber**

## Biotransformations in Organic Chemistry - A Textbook

2nd, completely rev. ed. 1995. X, 356 pp. 34 figs., 254 schemes, 25 tabs. ISBN 3-540-58503-6

The setup of this first textbook on biocatalysis is based on a professional reference book.
It provides a basis for undergraduate and graduate courses in biocatalysis as well as a condensed introduction into this field for scientists and professionals. After a basic introduction into the use of biocatalysts, the different types of reactions are explained according to the "reaction principle".

**K. Schügerl**

## Solvent Extraction in Biotechnology

### Recovery of Primary and Secondary Metabolites

1994. VIII, 213 pp. 130 figs., 48 tabs. ISBN 3-540-57694-0

The author details the reaction engineering principles and describes the recovery of low-molecular metabolites.
Besides practical examples for the recovery of different metabolites as well as for the calculation of the extraction processes necessary for equipment design, he also presents solvent extraction, novel separation techniques with liquid membrane, microemulsion and reversed micelles.

Springer

Tm.BA95.04.06

# Advances in Biochemical Engineering / Biotechnology

A. Fiechter (Ed.)

## Vol. 52 Microbial and Enzymatic Bioproducts

1995. Approx. 200 pp. 85 fig., 36 tabs. ISBN 3-540-59113-3

**Contents:** P. Rusin, H.L. Ehrlich: Developments in Microbial Leaching - Mechanisms of Manganese Solubilization.- S.Y. Lee, H.N. Chang: Production of Polyhydroxyalkanoic Acid.- S.A. Markov, M.J. Bazin, D.O. Hall: The Potential of Using Cyanobacteria in Photobioreactors for Hydrogen Production.- A.L. Gutman, M. Shapira: Synthetic Applications of Enzymatic Reactions in Organic Solvents.- Y. Inada, A. Matsushima, M. Hiroto, H. Nishimura, Y. Kodera: Chemical Modification of Proteins with Polyethylene Glycols.- F. Kawai: Breakdown of Plastics and Polymers by Microorganisms.

## Vol. 51 Biotechnics/Wastewater

1994. IX, 158 pp. 43 figs., 17 tabs. ISBN 3-540-57319-4

**Contents:** A.J. McLoughlin: Controlled Release of Immobilized Cells as a Strategy to Regulate Ecological Competence of Inocula.- A. Singh, R.Ch. Kuhad, V. Sahai, P. Ghosh: Evaluation of Biomass.- L.C. Lievense, K. van't Riet: Convective Drying of Bacteria II. Factors Influencing Survival.- S. Hasegawa, K. Shimizu: Noninferior Periodic Operation of Bioreactor Systems.- U. Wiesmann: Biological Nitrogen Removal from Wastewater.

Springer

Prices are subject to change without notice

Tm.BA95.04.06

# Springer-Verlag
# and the Environment

We at Springer-Verlag firmly believe that an international science publisher has a special obligation to the environment, and our corporate policies consistently reflect this conviction.

We also expect our business partners – paper mills, printers, packaging manufacturers, etc. – to commit themselves to using environmentally friendly materials and production processes.

The paper in this book is made from low- or no-chlorine pulp and is acid free, in conformance with international standards for paper permanency.

Printing: Saladruck, Berlin
Binding: Buchbinderei Lüderitz & Bauer, Berlin

11-2-95